MATH
RENAISSANCE

Growing Math Circles, Changing
Classrooms, and Creating
Sustainable Math Education

Rachel and Rodi Steinig

IS WHAT YOU MAKE OF IT!

Natural Math

Title: Math Renaissance: Growing Math Circles, Changing Classrooms, and
Creating Sustainable Math Education

ISBN: 978-1-945899-04-1
Library of Congress Control Number: 2017949994

Text: Rachel Steinig, Rodi Steinig
Editing: Karla Lant, Maria Droujkova
Illustrations and cover: Mark Gonyea
Layout: Jana Rade
Year: 2018

Published by Delta Stream Media, an imprint of Natural Math
309 Silvercliff Trail, Cary, NC, USA

Natural Math

Table of Contents

Introduction

Why do we have to learn math?
Why do so many people hate math?
Is memorizing an algorithm for a test really math?
If not, then what is?

I remember sitting at my desk in math class, idly doodling on my arm while the teacher droned on in the front of the room. As my head started to droop, I thought to myself, "Why do I have to be here?"

Surely you can relate. Most people I've talked to have a story about how math was ruined for them. A few people do say, "I loved math! It was my favorite subject"—like my mom—but more people talk about not understanding math, not thinking it was useful, being bored in math class, and crying over hard math problems.

So, back to the questions. What is this subject that so many people hate, and why do we have to learn it? And the biggest question: how can we improve math so that everyone can have better experiences learning it? Is it even possible?

This book is our answer to all of these questions. It is a compilation of anecdotes from the field, help with the pedagogy, and stories from people of all backgrounds. From mathematicians and kids, from teachers and parents. There are stories of success and failure, of love and of hate. Everyone has a story about their relationship with math. Here I am the scribe, the witness to the powerful bursts of emotion that accompany talking about math. Here people talk about standardized tests, human rights, learning math through dance, math geek stereotypes, inspiring teachers, and aha moments.

These stories can help us work towards long-term, systemic change in the way we teach math. It will take bigger social changes for this vision to fully realize itself, because math is a reflection of the larger society. But we don't have to wait; we can make a big difference in our class, family, and local community. Let's create a system where everyone knows what math really is. Where everyone is intellectually stimulated in math class. Where everyone has fun learning math. Where no one is afraid of math any more.

I want to see a golden age of mathematics, where everyone can access the beauty and joy in mathematics, a Math Renaissance, if you will. That might sound surprising. Isn't a renaissance a revival of art and literature, a time to appreciate the beauty of architecture and music? Many people I meet admit that the word "beauty" doesn't really come to mind when they think about mathematics.

But here's the thing: the ideas of "math" and "beauty" shouldn't be contradictory. Math can be beautiful. Math can be captivating. Math can invite us to ponder life itself.

So. A Math Renaissance. A revival of the practices that really work, and a transformation of those that can use improving. An age where everyone, regardless of experiences, gender, race, class, or anything else can see the joy in mathematics. See the beauty. Understand the essence of this subject that is so often misunderstood. Ponder deep mathematical questions and mysteries in collaborative groups. Feel confident in their math abilities. And understand what math is and why this subject is so important.

So, read on, and let's work together. Let's change the way children are taught. Let's bring back the joy into mathematics. Let's eradicate fear and foster creativity.

And, by doing this, let's change the world.

Rachel Steinig

Glossary

Rachel age 17 and Rodi

Mathematical thinking: Why does this mathematical concept play out the way it does? What assumptions are we making? How are these seemingly different math concepts actually the same? What would it take to break this pattern? Is it really a pattern? What is the underlying structure of this idea? What are other ways to express this mathematical concept? Mathematical thinking addresses the "why" of mathematics. In contrast, procedural questions only address the "how," such as, "How do you do long division?" or, "How do you graph the function y=tan(x)?" These are important questions: if you can't do long division you could lose the forest for the trees, stuck on a routine calculation while investigating a bigger problem; if you can't graph functions, you lose the chance to have a visual understanding of an abstract idea. When you delve into mathematical thinking, you also ask, "Why does long division work?" and, "Why does the graph of y=tan(x) look the way it does?"

Algorithms and cookbook math: We'll be the first to admit that we don't want to derive the quadratic formula or reason out the Pythagorean Theorem every single time we need to use them.

That's where algorithms come in, and why they are useful: they save time and energy. An algorithm is a recipe for solving a type of math problem. Good algorithms are clear, reliable, and efficient. "Cookbook math" is a term for a particular danger: over-reliance on algorithms. When students define a function as, "something that passes the vertical line test," don't question "negative twenty" as the answer to "How many people?" or stop asking "Why?" altogether, it probably means they were given algorithms before any deep mathematical thinking. In contrast, an inquiry-based approach invites you to derive algorithms, to explore their applications, and to comprehend them.

Math circle: A math circle is a community of people who get together to explore interesting math problems, projects, or topics. Various models of math circles exist, but they all have one thing in common: the desire to seek joy from mathematics. Our math circle is collaborative and inquiry-based. According to educator and math circle parent Melissa Church, "Traditional math teaching does not associate math with words, spoken or written. So this is a major shift in thinking. As long as the kids are discussing, questioning, arguing, and positing, then an important math circle objective is already being met. This emphasis on process over product is a—and maybe the—major distinction between math circle teaching and traditional math teaching."

Mathematical collaboration: Many people think math is an individual sport, not a team sport. In the real world of mathematics, however, mathematicians work together, and one person's progress is dependent upon someone else's work. This can easily be observed in problems that require a long chain of reasoning (such as proofs or logic problems); one person's statement prompts the next person's idea, and so on.

Accessible mystery: An accessible mystery is a problem that you understand and just can't help wanting to ponder—even if you think you don't like math or don't have time! It's that compelling. An accessible mystery is low-floor, high-ceiling: easy for all to enter, and with limitless potential for deep thinking. A good example is The Monty Hall Problem. Go ahead, search the web for it!

Intrinsic motivation: This happens when your own hopes, dreams, and causes propel your learning. For example, you learn how to drive not because you're going to get an A on the test, but because you want to be safe, comfortable, and efficient driving around. Your intrinsic motivators may be curiosity, choice, or growth. This is in contrast to extrinsic motivators such as grades, gold stars, or punishments. Intrinsic motivation for learning works long-term. It causes people to put in more effort and to process information more deeply. Extrinsic

motivators can help a child push through a rough patch or add game elements to a boring task. Unfortunately, extrinsic motivation can undermine intrinsic motivation. To avoid that danger, plan few tasks that are too scary or too boring (listen to the students during planning!), so you will need extrinsic motivators sparingly.

Learning modalities: When Rodi went to dancing school as a child, she couldn't memorize the dance routines by watching the teacher and other dancers, or by just repeating the steps herself over and over. She had to have a rhythmic chant in her head with the names of the steps. She learned through the combination of spoken words and rhythm. Different people learn via different modalities: visual, auditory, kinesthetic, and tactile. Most people need to explore mathematical concepts via multiple modalities: seeing a graph, hearing about applications, making a hands-on model, and so on.

Mistakes: Don't fear them. In fact, embrace them, because you learn from them. How? By tracing mathematical thinking that directed you where you ended up. When students make a mistake, you can wait for them to discover and correct it themselves. If you see a train wreck coming, you can offer gentle guidance, such as, "Can you explain your reasoning in that last step?" Of course, first think to yourself, "Could some good come of me letting that train wreck happen?" and "Would it even be a train wreck?" If you correct student mistakes too often, they can grow dependent and lose their problem-solving confidence. If you try too hard to prevent the train wreck, you might engage in rescuing. Do you often find your students on the brink of their breaking point, where you have the urge to rescue them out of compassion? If so, tweak your style. Stay attuned to students' emotional needs, encourage them as you go along to maintain high hopes, and re-frame the same question in many ways so they can eventually tackle it.

Mindset: Mindset refers to your attitude about something, and how it affects your experience. When you're driving to work in traffic, you have a choice. You can tell yourself, "I'm going to be inching along the highway, grumble grumble grumble." Or you can tell yourself, "Finally! Here's my chance to listen to my favorite news show and get caught up on what's going on." Either way, it takes the same amount of time to get to work. But which was a better experience? Mindset not only affects attitudes and perception of experiences, but also brain growth, and your math performance. If you believe that brain growth is possible, it is more likely to happen. If you believe you can improve your math performance, you are more likely to do just that.

Struggle: Struggle is a good thing. It makes your brain grow. This is in contrast to drudgery, which people sometimes equate with struggle. For people who fear math, struggle can cause anxiety. Also, too much struggle can lead to burnout or getting so stuck that they need rescuing. So struggle is good in moderation. Once Rachel worked on a geometry proof every day for a week; finally she finished it—a good struggle! It made her feel challenged and accomplished. When you're in a class above your level and can't understand a single thing, that's usually too much struggle, making you feel lost and hopeless.

Sometimes people also equate struggle with frustration, which is inevitable in math. Frustration can get in the way of mathematical noticing, and the lack of noticing can get in the way of solving problems. You are not born with a set level of frustration tolerance. It can be developed. It's important to tell students this. It's also important to acknowledge their feeling of frustration when it inevitably bubbles up. When you acknowledge how a math problem makes you feel, it is a relief and a release. It frees up your brain into solving the problem instead of fighting emotional reactions.

Brain growth: Brain growth comes from forming new neural pathways and pruning old ones, which makes you a stronger thinker. Recent developments in neuroscience have revealed that our brains do not stop growing, ever. They always retain some plasticity. Your brain grows when you engage in mathematical thinking: when you pose questions, when you struggle, when you seek your own ways, and when you fail. Your brain grows a lot more doing those activities than when you use a known algorithm to solve an easy exercise.

Metacognition: If you're thinking about your thinking, you're engaging in metacognition. For example, "How did I just figure that out?" or, "Why did this explanation make sense to me?" It's not enough to engage students in problem-solving and hope the strategies soak into their long-term memories. Name the strategies for what they are. Better yet, ask the kids, "What strategy did you use to figure that out?" Keep a running list of strategies on a board, readily accessible for everyone. Soon you'll notice students looking at the board without your guidance, discussing which strategy worked best, and developing a taste for favorite strategies. Metacognition promotes problem-solving skills such as productive coping with frustration, self-reliance, and confidence (Safari and Meshkini).

Student-directed learning: Do students have much say in what they learn and how they learn it? The goal of student-directed learning is to give each student more control, more ownership, and more responsibility regarding their own education. That can lead to a more effective and meaningful education. For example, a student said, "I want to understand the relationship between the times tables, exponents, and dimensions." Definitely not in the textbook. This student learns best when she sees how things are interrelated. So we got out some wooden cubes. She was delighted to explore the areas of squares (side times side), volumes of cubes (multiplying three sides, or the third power), and then the higher dimensions and higher exponents. Later these topics came up in a math circle. Another student stated "those topics have nothing to do with each other." The first student was able to say, "Actually they do, and here's how." Her understanding of the concepts that were the purpose of the lesson (and the math circle) was strengthened because she had a say in her own learning.

Socratic dialogue and the Socratic method: In Socratic dialogues, at least one person plays the role of Socrates, the character who confesses little knowledge about the topic and asks many questions. The goal is for all participants to learn more, both about the topic and about their own perspectives. In the Talking Stick Math Circle, a facilitator's Socrates character throws out a question, and then everyone in the room bandies about ideas as if they are all Socrates. It's a collaborative exploration with everyone on an equal footing. (This is in contrast to the type of Socratic dialogue where the Socrates' character interrogates and confuses the others to reveal the flaws in their perspectives.) In our math circle, open-ended questions play a big role. The facilitator doesn't try to lead students to converge on a predetermined result.

Inquiry-based education: An approach where facilitators pose questions that are not leading in a predetermined direction, thereby giving students ownership. The goal is for students to discover things themselves by then posing their own questions. So a facilitator might ask, "What do you wonder about proofs?" The students would then respond with their own more specific questions and plans to get them answered. The facilitator is not a teller of answers. Dialogue, problem-solving, and mystery are tools of inquiry.

Suppose you were working with your child on the equation $6x + 3 = 7x$. If you ask, "What is a good first step?" the student might answer, "Isolate the variable," or "Combine like terms." The unspoken assumptions here are that we are trying to solve for x, and recall the pre-determined methods of doing so. Inquiry-based mathematics casts a wider net, or maybe no net at all. You might ask "What's interesting about this?" or "What's puzzling here?" or "Do we know anything about this equation?" The students might respond with, "We can tell that $7x$ is bigger than $6x$" and lead into a discussion of whether that's always true. Or even, "I wonder why the variable is traditionally called x," leading into an investigation of math history.

Coercive education: A pedagogical approach in which what students learn and how they learn it is dictated to them. In Psychology Today, Peter Gray discusses coercive education: "[Children have] instincts to explore; to observe; to eavesdrop on the conversations of their elders; to ask countless questions; and to play with the artifacts, ideas, and skills of the culture all serve the purpose of education ... Children's natural means of education require freedom ... Children crave learning, but they crave it on their terms. They learn well when they are in control, and, like you and me, they often become resentful when others try to control them." Coercive education ingrains people to jump to tasks someone else requires of them, quickly getting the right answers by the given methods, or else. "Or else" may mean embarrassment, lower grades, or disappointment of parents and teachers; the whole endeavor is driven by fear of mistakes. Even when there is no outward pressure anymore, such as in the context of a math circle, the participants may continue to behave as if they are in a coercive situation. For example, facilitators wrestle with the temptations to tell students exactly what to do. Even children may impose "the one correct way of solving problems" on one another. It takes a while to shift the culture away from such practices. Be patient with yourself and others.

Frustration: Frustration is inevitable in math. One goal of math circle is to better handle frustration, which can get in the way of noticing. You are not born with a set level of frustration tolerance. It can be developed. It's important to tell students this. It's also important to acknowledge frustration when it inevitably bubbles up.

Math history: What motivated people to work on questions such as, "How do you divide an angle into three equal parts? What were the implications of math breakthroughs for the society? How did people react when popular conjectures were proven wrong? What were the life stories of the people involved? How did people figure general patterns out before variables were invented?" We humans like to understand things in context, not in isolation. So it's empowering for students to learn the history behind mathematical principles. When they hear the stories, they can see themselves as the actors within, making the math come alive.

Math philosophy: Do mathematical ideas have to be grounded in physical reality, or can we just make stuff up as long as it doesn't violate its own logic? Would math exist if humans didn't exist? What is required for proof? What makes one proof more beautiful than another? Is there a difference between truth and fact? How should we think? For instance, should we collect a bunch of data and then make generalizations about what we see (induction)? Or should we start with general principles that we already know and then make logical deductions? Is there math in everything? What's the role of intuition in mathematics? Is math invented or discovered? Philosophy of mathematics tackles these fundamental questions, and extends them into the larger picture. Studying the philosophy of mathematics impacts how we enjoy math classes—and also, the way we live our lives.

What is Math, and What Does It Mean for You?

Rodi

This book is not a set of instructions.

That's because math itself is not a set of instructions. If someone instructs, "When you see something that looks like this, do A, then B, then C," and you do it, you are not doing math. You are just following directions.

I must admit that I like following directions. My favorite math course as a student was Differential Equations. In the DE course I took, students had two tasks: (1) to examine a given equation and recognize its general form, and then (2) to apply the appropriate algorithm (predetermined set of steps) to solve it. I felt the same sense of accomplishment from solving those problems that I feel when I clean my car.

When I learned much later in life that not much mathematical thinking had been happening in that type of math exercise, I worried that maybe I didn't actually like mathematics itself. Was real mathematics esoteric, difficult, and only available to high-level specialists? Was it like designing and building a car from scratch, without instructions?

Before the days of GPS, I spent many hours in my car driving around different neighborhoods on my way to my many teaching destinations. In my trunk, I kept a book of maps for every nearby county in case I got hopelessly lost and might be late. But I loved to keep that book locked away in my trunk so that I could discover different routes through experimentation, a little bit lost on the way. Math feels more like that: like driving around or wandering on foot, available to every child, though you can get very sophisticated with maps, GPS, and satellites. It's doable by regular people.

Two questions arise from these reflections:

If you are not engaging in mathematics when you follow instructions, then when are you doing math? And, if this book is not a set of instructions, then what is it? It's a set of suggested ideas and questions. I posit1 that asking questions is the root of mathematics. "Posit" means to put forth an idea that might actually be right, but could be wrong.

One way to look at math is the search for patterns and the simultaneous debunking of seeming patterns. In other words, math is about making conjectures and seeking counterexamples. American math circle pioneers Bob and Ellen Kaplan refer to it as a pursuit of humanity's architectural instinct. All humans have an innate desire to seek the underlying structure of things. Once I led a math circle course based upon a narrative story involving sorcerers. Some sorcerers were male; some were female. By the second session, all of the students were putting a lot of energy into ascertaining a pattern. Surely it must be a predictable sequence of genders. Surely they could predict the gender of the final sorcerer that would arrive by class six. In fact, it was random. I had never mentioned, hinted at, or consciously created such a pattern.

Paul Lockhart states that mathematics is "the art of explanation … the argument." It is the essence of math to examine such questions as whether there's a sorcerer pattern, how you can describe it, and how to make predictions from it. In other words, mathematics is more the act of turning the problem on your mental lathe versus grabbing the answer (which may not even exist, as is the case with the sorcerers). Math is imagining, designing, or making your own bowl rather than buying one from a store.

In my students, I see that human instinct to seek structure. They use mathematics to understand our world—and to make up their own patterns when things seem random or chaotic. According to author Salman Rushdie, we use stories in the same way. In his memoir *Joseph Anton*, he states that stories "could make him feel and know truths that the truth could not tell him," and that a human was always a creature who "told itself stories to understand what kind of creature it was." So, do we also do math to understand what kind of creature we are?

In my math circles, we use stories as a rampway into deep mathematics. Are we using those stories to understand math to understand ourselves and our world? Are storytelling and mathematics two sides of the same coin?

This book will give you ideas about how to intertwine stories and mathematics. Try to posit your own definition of mathematics. If my students say, "I posit that…" I feel like I've facilitated some mathematical thinking. I say "posit" once again because mathematical thinking is open to change, to new ways of defining things. Two plus two doesn't always equal four; we could be working in base three (where 2+2=11). Talk to your children about what they think math is. At first, they (and you, and many other grown-ups, even teachers) may say mathematics is counting, adding, multiplying: a performance of number skills. Follow stories, problems, and accessible mysteries in this book and beyond. Keep asking yourself and your children what math is. My hope is that a whole new worldview, or even new world, opens up to those who move beyond the arithmetic.

I like to think of math as figuring out what's really going on, and what it means. The purpose of my math circle and this book is not for students to become better at memorizing and applying algorithms. (There are plenty of excellent books on that topic.) My goal here is for students to feel that they have the power to create their own algorithms if they want. My hope is that students will be sufficiently trained in mathematical thinking to ask themselves, "Why does that happen?" and will therefore be able to struggle to the answer. Struggle is a good thing. More about that later.

Along the way, we will delve into the history and philosophy of math—two topics that make our explorations more meaningful for the students.

I am grateful to have the good fortune to facilitate math development in a noncoercive, inquiry-based classroom. Many teachers have it hard with requirements to teach particular content in a proscribed way. The good news is that there are a lot of little practices you can implement in any environment to move towards the freedom that mathematics celebrates. This book has a lot of pieces of math circle culture that you can adopt.

Also, you don't need to have taken such high-level courses as Differential Equations to use this book. In fact, to facilitate mathematical thinking, you don't even need to know the answers to the questions you pose. Your exploration can be even better when you don't! Not-knowing can bring more authentic inquiry to your discussions, because your expectations are less likely to direct the inquiry.

Whether you are in a classroom, living room, or math circle, my goal is to inspire you to question everything with your kids.

Rodi Steinig

Assumptions About Math

Rachel age 15

What do we assume about math? If you really think about it, there are lots of things we assume:

- Little kids can't understand complicated math topics and ideas
- People who are good at math are geeks, wear glasses, are antisocial, etc.
- There's only one way to do math
- You can't change your math performance
- Math is super boring

Now, I'm going to go through all of these assumptions and cast doubt on them.

Little kids can't understand complicated math topics and ideas. I've observed countless math circles that my mom has taught to little kids. They are *so* adorable! But anyway, she's helped them discover some pretty amazing things, things that I didn't even know.

For example, in one math circle the kids were working on a problem involving saving a dying unicorn. There were Harry Potter characters that all had to cross a bridge safely under a certain amount of time to save the unicorn. However, all of the different characters walked at different speeds. The kids pretended to be the characters and walked across the bridge together. They discovered that when two moving objects are attached together, and they move at different rates, they have to move at the rate of the slower one. This is a huge conceptual leap for children. Now, at the time, *I* didn't even know this. This is an example of how young children can understand complex mathematical concepts when the concepts are presented in an accessible way.

Young kids are even learning calculus! My mom just finished a five week math circle with 5-6 year olds. They did some calculus and discussed whether infinity is linear or circular. Is traveling around the earth around and around forever mathematically the same as traveling into outer space on a rocket ship? Now, just to point out, these kids are average 5-6 year olds, not "geniuses." Math can be accessible to anyone, regardless of their age.

Now, the next assumption: **People who are good at math are geeks, wear glasses, and are antisocial.** There are a lot of stereotypes about people who are good at math. Most people assume that "math people" are skinny, sunlight-starved, male, Asian, glasses-wearing, antisocial geeks who stay inside and study all day. These are just a few stereotypes. I'm sure you could name more.

Under scrutiny, though, we can see how obviously ridiculous these stereotypes are. Okay, listen up: this needs to be said. There isn't one type of person who is good at math. Not all people who are good at math or just enjoy it are anti-social, or unathletic, or Asian, or wear glasses. For example, Danica McKeller was a TV star before becoming a mathematician. Field Medal winner and mathematician Maryam Mirzakhani is of Iranian descent. African-American pro football player John Urschel is also a mathematician pursuing his PhD at MIT. You can be great at math and have tons of friends. You can be great at math and wear pink, read *Teen Vogue,* be Black or Latino, female or male, fat or thin, or poor or rich. Anyone can be great at math.

Some of these stereotypes might seem funny, but stereotyping can have serious consequences for kids. Children are often bullied for being good at or interested in math. This is terrible, not only because it's cruel, but also because it makes kids avoid math because they are afraid they will be teased.

Hopefully someday children will not be teased for being good at things that seem less socially desirable. In a perfect world, math and all subjects would be taught in ways that allowed kids to enjoy learning whatever interested them most. Under these circumstances, I don't think the "uncool" stigma would attach to math. I think the first step is to create an open and supportive environment where children feel comfortable excelling at the subjects they love.

Next assumption: **there's only one way to do math**. Now, what I love about math is that it's an art. It's like music. The patterns, lines, and variables harmonize with each other to create beautiful symphonies. Math is an art because there are *tons* of ways to solve a problem. This becomes really apparent in algebra. In an algebraic equation, with numbers and variables on each side of the equal sign, you can rearrange them anyway you like. It doesn't matter which variables you add and subtract in which order.

What some people love about math is the order and structure—that 5+5 will always = 10. The artistic freedom of math doesn't negate its appealing order. There are still rules that you have to follow. If you are using base eight, 5+5 will not = 10. And even in base ten, where 5+5 will always = 10, there are many ways to find that answer. What's wonderful about math is that there is structure, but freedom as well.

Next: **You can't change your math performance.** I bet you've heard someone say "I'm not a math person," or have even said it yourself. But guess what? This can change. *You* are in charge of your math fate; how hard you work can change your math performance, as well as your mindset.

In her study, "Mindsets Matter," Jeni Burnette and her colleagues found many negative correlations between helplessness-orientation and achievement, but many positive correlations between mastery-orientation and achievement (Burnette et al.). Helpless-orientation means believing that you can't change your math performance because intelligence is a fixed, unchanging quality, and mastery-orientation means believing that you can change intelligence with applied work. The results of this study show that your expectations can create self-fulfilling prophesies. A recent study even suggested that believing that you can't change your math performance might be responsible for a lot of the gender gap in mathematics (Ganley & Lubienski; Gholipour; Cimpan et al.).

The issue here isn't whether your math-related abilities such as short-term memory or spatial reasoning are controlled by your genes, or how "good" your genes are in this area, because no one knows for sure which genes are involved or to what extent. This remains a significantly controversial area. However, scientists *do* know for sure that you can improve your math performance through hard work and changing your mindset.

Now, onto the last assumption. **Math is super boring**. The reality is that math, like anything else, can be boring—but it doesn't have to be. Let me tell you a little bit of my story.

I was homeschooled my whole life and I started attending math circles in middle school. Before I went to math circles, I didn't like math that much. It was okay, but I got really frustrated when it got hard. It's embarrassing to admit, but I even threw stuff at my mom.

When I started going to math circles, I was like, "Is this *real* math?!" My mom said it was. I was astonished because the math that we did in math circle was really fun, unlike the arithmetic that I had done at home. I was pretty surprised. Going to math circle made me love and really appreciate math. Before, I thought that math was arithmetic, because arithmetic was the only kind of math I had ever done. I still hate arithmetic. Who has time to learn all of those times tables?! In math circle, I was exposed to lots of different types of math, like geometry, probability, statistics, graph theory, compass art, compass and straightedge constructions, and Euclidean geometry. And guess what? I loved them all!

A few years ago I started as a 9th grader at a public magnet high school for high-achieving kids. It was in a school district that was and is broke, in one of the poorest big cities in the country. When I started in 9th grade, let's just say that math was far from my favorite subject. It was my most boring class. We learned the algorithm for something and then we did a lot of practice problems and then we had a test on it. It wasn't hard—in fact, it was easy—but it was really boring.

So I've experienced times when math was hard and frustrating, times when it was amazing, and times when it was boring. I've learned that math can be really fun if it's taught in a fun and interesting way, such as in math circle. Well, in math circle it wasn't really taught, we just discovered it for ourselves. I think self-discovery is a good way to learn, because you have fun discovering things for yourself. In public schools, math is hardly ever taught this way; it's mostly taught by lecture and rote memorization.

I also think that the type of math you are learning is crucial to your enjoyment. For example, I don't like arithmetic, but I love lots of other types of math. There are sixty different types of mathematics. Sixty! Everyone likes different things. No one is going to like every kind.

It's important that children get exposed to different types of mathematics, so they find something that they like. And I'm not suggesting something crazy, like that kids have to learn all sixty types of math. But right now, arithmetic is just about the only type of math taught in school in the lower grades. Kids aren't exposed to any other types of math, and they get the idea that arithmetic is the only type of math. If they don't like arithmetic, they think that they don't like math. They have no idea that there are fifty-nine other types of math out there waiting to be discovered.

I wonder if arithmetic/computation should be the first kind of math taught in school. Maybe if another type of math was taught first, or a mixture of many different types of math, kids would grow up liking math more. Everyone likes different types of math, and it's important that you get exposed to all different types to have a full appreciation of the subject.

I've demonstrated that math isn't all drudgery and boredom, but I realize that I haven't really proven that arithmetic isn't boring. I had terrible experiences with arithmetic, so it's hard for me to see it as fun. I didn't memorize anything because when my mom tried to teach me, I just threw stuff at her.

I learned arithmetic on a need-to-know basis; when I needed to know a certain math skill to do something I wanted to do (shop, cook, or analyze data for science fair projects), I learned it. There's research backing up up this way of learning. "In the 1995 School Restructuring Study, conducted at the Center on Organization and Restructuring of Schools by Fred Newmann and colleagues at the University of Wisconsin, 2,128 students in twenty-three schools were found to have significantly higher achievement on challenging tasks when they were taught with inquiry-based teaching, showing that involvement leads to understanding. These practices were found to have a more significant impact on student performance than any other variable, including student background and prior achievement" (Barron & Darling-Hammond). Of course, there will be things that are prerequisites for other types of learning that won't come up on a need-to-know basis, and you will need to go out of your way to acquire this knowledge.

This makes me wonder: are there ways to make arithmetic and other more technical grinds behind mathematics fun? Of course, some people naturally like arithmetic, but for those who don't, how can it be made bearable? Fluid skills help with math, but many children dislike practicing.

There are ways! For example, my mom teaches arithmetic through unsolved questions in mathematics. Kids try to solve math problems that no one has ever solved before. There's a lot of incentive because, well, how cool would it be to solve a math problem that has stumped mathematicians for years? And, if you win, there's a prize of one million dollars. How awesome is that?

Also, another possible way to make arithmetic fun is to learn it through moving your body. You can discover all you need to know about negative numbers by jumping on a sidewalk chalk number line. Malke Rosenfeld is a big leader in this field and does body math via dance.

Unsolved problems and movement are just two of many techniques for making arithmetic make sense to you; there is always another option. You can learn arithmetic skills in the context of something you want to do, be it million-dollar math problems, movement, or maybe robotics, storytelling, or crochet. This way, you have immediate motivation to practice it until you learn it well.

Compass Art

Rodi

It is said that a little bit of stage fright is a good thing. That's good to know.

I was a little nervous about leading a math circle for young middle schoolers (ages 9-11, grades 4-6), despite having worked with that age group in various settings over the years.

Worry #1: Would they ask me questions I couldn't answer? I've been comfortable saying "I don't know" to students of all ages for decades—why worry about this now? A bigger worry, though, was

A little bit of stage fright means you care enough to do a good job. It gives you an adrenaline boost to enhance your performance. Teaching is at least partially a performance. Hence, if you don't have stage fright, well, you can fill in the dots…

Worry #2: Would they find the material uninteresting?

This was only the second math circle I had ever led. The first had been a rousing success, with younger kids and an accessible mystery. I didn't have an accessible mystery here, or even a question. That was probably the source of my apprehension.

One of my mentors, Amanda Serenevy, said that compass art would be sure to capture their fancy, and mine, no matter what.

She was right.

I showed the kids some compass art, asked them some general questions about it, gave them compasses to play with, and the math flowed from there. We met weekly for an hour for six weeks. What blew me away is that over the weeks, the students ended up devising their own accessible mystery.

This is the story of the evolution of that accessible mystery over the course of six weeks. The mystery evolved, as did my math circle facilitation skills and confidence, through questions asked by me and the kids. I'll tell the story through those questions.

WEEK 1

At the start of week 1, I covered the table with various materials:

- images by Michaelangelo;
- images by Bernini;
- images by Lloyd Wright (or, the students wondered, was it simply Wright?);
- images by Zarah Hussain;
- Native American geometry designs;
- art created by mapmakers;
- the Feng Shui compass;
- and a photo of a piece of flower-of-life jewelry.

I asked the first question:

What do all of these things have in common?
After many conjectures, including octagons, circles, and symmetry, R said, "You can make them all with a compass."

What is a compass?
We had to clarify that the compass we mean here is the tool used to draw circles, and then,

Who do you think invented the compass?
No one was sure, but Z suspected Galileo. After we talked about Galileo's famous accomplishments and his pretty awesome full name (Galileo Galilei,

which only A knew), I recounted an anecdote about his sour experience in medical school lectures. Then I showed them a painting of Euclid holding a compass 1,800 years before Galileo was born, and told the story from Greek mythology about Daedalus' assistant Perdix and the invention of the compass. I told them that Wikipedia says that Galileo invented it. "You can't trust everything you read there," said the kids. I told them that it seems certain that Galileo made some significant improvements in the compass, but that I am still trying to find an authoritative source on its true provenance.

The kids were eyeing some compasses in a box on the table, so I asked,

According to Reimer's Historical Connections in Mathematics Volume I, "When Galileo was 17, his parents sent him to the University of Pisa to study medicine. The young student found the lectures boring; there were no opportunities to experiment or work in a laboratory. One day while walking down the university's hall, he overheard an excited professor discussing geometry. As he peeked into the classroom, he was surprised by the involvement and interest of the students. Here was a fascinating subject! Much to his father's chagrin, Galileo gave up the pursuit of medicine and devoted his life to mathematics and astronomy." We discussed G's comment that lecturing is the least efficient form of instruction.

Would you like to use these compasses?
A resounding "Yes!" from the chorus. We played with them for a while, then looked more closely at the picture of the piece of jewelry. It was a gold "flower of life" pendant—a geometric design with history in many cultures and religions. The students struggled with how to draw it. They debated their own question:

Is the flower of life constructed with any straight lines?
Discourse and experimentation continued. J came very close, with a similar center, but a more rectangular outward flow, almost like a compass rose. We traded compasses so that everyone could try and critique each type. The students studied that photo of the necklace again and again, trying to come up with strategies on how to make the design.

As everyone worked, we looked more at the work of Islamic artist Zarah Hussain. I told of her mathematical studies, of the goals of her art (to understand her religion), and of the great influence of mathematics on Islamic culture and religion. I mentioned that in this culture, the circle was once used as a unit of measure. The students were curious:

How could a circle be used as a unit of measure?

J suggested that perhaps a rope was used. I replied that his comment reminded me of a photo I had printed out: a dog tied on a rope in a yard, producing a circular pattern pressed into the grass as it ran.

This discussion led into Euclidian constructions, also called "compass and straightedge constructions," geometric figures that are constructed solely with a compass and a straightedge (no ruler!). Everyone wondered,

Is it possible to make every geometric figure with a compass and straightedge?

We briefly discussed this, but by now what students really wanted was a hint about how to make the flower of life, so I gave them one.

As the designs blossomed into flowers, G still wasn't convinced that the pattern had no straight lines. I asked,

Since you mention it, what is a line anyway?

We had a hearty conversation about just what is the definition of a line. I was leaning pretty hard on James Tanton's book *Geometry: Volume I* here, and all throughout this course. In preparation for the course, I had read the book, and came to class armed with my edition of the book, full of Post-its.

1) Draw a circle. Lock your compass at this radius.

2) Place the metal tip of your compass on the circle's circumference. Draw a new circle of the same radius.

3) Draw a circle at each of the points where the two circles intersect.

4) Repeat step 3 until you have six circles surrounding your original one.

5) Repeat step 3 as long as you want!

The students attempted to define "line" with the word "straight." I asked what "straight" meant. They ended up defining it with the word "line." It was impossible to define "line" without using circular reasoning, a concept that everyone found fascinating—and a bit frustrating. Z pled:

So tell us your definition of straight!?

A line is considered undefined in math, I said, but we all agree to use the term with an understanding of what we mean. We talked about how, in math, we choose to believe some things even though we can't know them for sure. For instance, how can we really know that a line goes on forever, or even, suggested Z, that something can be infinite? The students went home that week with the intention of trying to come up with their own definitions of a circle.

WEEK 2

"I worked on my Flower of Life at home," said Z as the students entered the room the following week. As she showed me her work, the others began to draw compass designs long before our circle officially began.

A formally began the second session by asking:

So, what is the definition of a circle?

The children were excited to discuss the circle definition that Z had brought in. The definition went on the board, and then underwent intense scrutiny for circular reasoning. None was found, but the terms in the definition needed clarification. Some terms were readily defined by the group: *continuous*, *defined*, *diameter*, and *circumference*.

(A modern definition of circle is something like, "all points equidistant from another point." Euclid said, "a plane figure contained by one line such that all the straight lines falling upon it from one point among those lying within the figure equal one another.")

Radius was a tough one. I pushed hard for a definition that did not use the words diameter or line. R finally came through with "line segment" and someone else suggested center, which moved us into the next debate:

Can the term "point" be defined?

Numerous attempts were made, as I read them commentaries from mathematical philosophy on definitions in general, and more specifically the evolving definition of a point. Euclid said a point is "that which has no part." Could you do better? The students bandied about more attempts—dot, sphere, disk, location—as I told more about Euclid and his attempts to define geometric terms. When I read them his definition of a line ("breadthless length"), A exclaimed, "What?!" and we all concurred. We concluded with a consensus that while a circle is definable, a line and a point are not.

The students experimented with Euclidian (compass and straightedge) constructions. They wanted to figure out:

How do you construct a triangle with a compass and straightedge?

They discovered how to use Euclid's method to construct triangles, and to make 6-pointed stars within the circle. Most of the kids wanted help in finding the points to connect to make a star-within-star design. I gave hints about using perpendicular and parallel lines within the design to guide them. Then the kids inscribed stars within stars until the design got too small or frustrating or until the thickness of their pencils prevented further progress.

We observed how infinity can go in more than one direction. You can keep getting bigger, for instance, or keep getting smaller. (Mathematical note: there are also other ways infinity can progress which are beyond the scope of this discussion.) A experimented with concentric circles. J began coloring his creation, exploring symmetry with color. G had seen a colored compass design in my notes that she liked. I told her that it was a Euclidian construction of a square (compass and straightedge only). She wanted to attempt this on her own with no hints. I went against my better judgment and gave her a hint anyway ("You might need to draw some arcs to make it, and here is what an arc is").

This hint may have undermined her creativity. I had one agenda in my mind (using arcs to draw squares) when in fact, who knows, maybe there is more than one way to do it? And if not, why steal her joy of discovering that for herself?

G cheerfully called my bluff by constructing an apparent square without the need for any arcs. Once again I was reminded of the human architectural instinct; we inherently want to discover the structure of things. And when we're given the freedom to explore, we do. We all do. We just do. (G and I then talked about how she could check for sure to see whether it's definitely a square. I wonder if she will.)

For the most part, I kept my "teach-ering" instinct in check. I let go of my attachment to any particular agenda by telling math history stories as the kids worked on their constructions. I told them about:

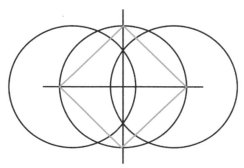

Euclidian construction of a square

- Euclid's famous reported quote to Ptolemy I ("There is no Royal Road to geometry,");
- the ancient library of Alexandria;
- and Galileo.

I had planned to talk about the beginning of his persecution, but the students knew something about the end of it and therefore discussed his trial and death. The students were aghast that it took so long (until 1992!) for the Church to officially pardon him. "Did he have no family trying to clear his name over the centuries?" "Did they forget?" (One of the many things I love about math circles is that I always leave filled with my own curiosities about all matters. The students' reaction to Galileo's overdue pardon made me wonder this: would people of other times have been as aghast as my students, or is that a phenomenon particular to our time, a time when the websites that clean up reputations do a huge amount of business? On the other hand, in the "olden days," wasn't family name everything? Time to visit the Internet…)

At this point you might be thinking I know a lot of history and math. You may be wondering, "How could I lead a math circle if I am unfamiliar with math history?" But, to pseudoquote one of my favorite literary characters, Junie B. Jones, "Guess what! I probably know way less history than you do." I am, though, extremely dexterous with the Internet, and have some good math history books. As far as the math goes, I knew a lot of math the way it was taught in my schools when I taught this course, but in math circles we're going deeper. We are covering a lot of the same mathematical content that is in the school curriculum, but we're doing it via mathematical thinking instead of the cookbook approach. I've learned a lot as we go along. When I facilitated this course, most of the time I felt like a student, too.

Even the masters don't know everything. There's just too much math—over 60 specialties. We all feel in over our heads. (Just imagine how kids must feel if this describes us, their doubt-filled leaders. According to Rachel, the knowledge that teachers don't know everything makes students feel both empowered and insecure—empowered for obvious reasons, but also a bit insecure that maybe the teacher doesn't know enough to impart wisdom. The question that naturally arises, then, is where is the balance? How much teacher knowledge/doubt is required for the kids to feel comfortable? Is there a threshold?)

To me, the math circle world feels like a guild of old. At first, I felt like an apprentice—live trainings, listservs, published material, professional organizations with resources, and so on. I submitted all of my work to the masters, using modern methods, i.e. publishing a blog about EVERY math circle I led that the masters read and commented on; unrelenting question postings to the listservs; email inquiries sent directly to the masters…I've experienced such a culture of cooperation in the math circle community. Every week, many people have come to my aid with suggestions, feedback, even corrections to my mathematical knowledge. This still goes on, but now people sometimes ask me for guidance. I think I've become a jour-neyman—still learning, but knowing enough to hold my own a little more than before.

WEEK 3

> When exploring topics that you don't feel total confidence in, you might tell yourself things like, "Well, at least I probably know more than the students. And if something comes up that I don't know, we can find out together as the students benefit from my modeling of inquiry skills."

I had set out photos of a Julia Set crop circle and the Cissbury Ring crop circle to whet people's appetites for circles. I knew almost none of the math behind these patterns, but figured that the kids would sense the math in them, and that we could discover it together if curiosity prompted us to.

Everyone had many questions about the crop circles:

What's the deal with crop circles? Specifically, how did they get there, and how did they get so perfect?

I said, "I don't know," and, "I don't know," and, "I don't know," again. (Remember, in this style of math circle, leaders do not present themselves in positions of content authority. This moment was a great chance for me to reinforce that with the kids.)

The kids gave many conjectures, and I answered with what little I know about them. Then the kids wanted to continue the process of defining terms that we started a few weeks ago. Each week, the process was becoming even more collaborative. We asked,

What is an angle?

With each attempted definition of "angle," the kids themselves automatically questioned the definition of terms within the definition, and refined their statements. R threw in some math vocabulary she knew. The group ended up somewhat satisfied with the definition "2 line segments coming out from a common vertex." Since this definition said nothing about measurements of angle and the students intuitively felt that the definition must somehow talk about "distance," they were not totally satisfied with it. (This is exactly the kind of real mathematical thinking that a math circle encourages. I let them know that they were thinking like mathematicians with all their questions and doubts.)

I asked them to consider defining an angle, in a less formal way, as a measurement of turning. I used 2 pencils to show a full circle of movement and asked, "how many degrees in a full turn?" "360," said several voices. "How about half a turn?" "180 degrees." "A quarter of a turn?" "90 degrees." Then I asked,

And what is a degree, and why do we use 360?

Now the voices were silent. Time for some math history. I started to tell them about:

- How this number originated with the Babylonians. "I love that name, Babylonians," said M, and we said it a few times. He knew that we use 360 because of something about the earth's rotations.
- I told them that it takes the earth approximately 365 days to travel around the sun. "Approximately?" asked Z.
- So then we talked about how leap years work before getting back to the ease of using 360 instead of 365.25 to measure a circle.
- I mentioned how the Babylonians looked at math from a practical point of view, as opposed to the later Greeks, who applied the rigors of reasoning to it.

Then I asked,

Do you see the practical problem of using 360 for a circle if we were to travel to some other planet?

M explained how another planet would have a different number of days for its orbit. I explained that mathematicians have tried to avoid that earth-centric approach by measuring angles in a unit other than the degree, such as the radian (based on a circle of radius 1), or the gradian (a measurement where ¼ turn is called 100 gradians). A seemed to really appreciate the idea that a number of things in math have been created, not discovered, by people. So I asked,

Most of the kids seemed to really like the gradian, except for M, who wanted ¼ turn to be called "1,000 Whatevers," so that even more accurate measurement would be possible. So I moved on to my next question,

"360," said at least 3 assured voices, while the others remained silent. Then I asked each child to draw a large triangle on a piece of paper, and follow its angles with a pencil. Each noticed that when the pencil returned to its starting point, it had undergone half a turn.

"360! Er, uh, 180!" They thought about what they had just done, and then all agreed that triangles actually have 180 degrees. Z, looking thoughtful and confused, asked,

The group then had a very brief conversation about the value of knowing why something works in addition to how to do it. (1) You want to be able to take the mathematical reasoning you use on one problem and apply it to something similar but not exactly the same. (2) You want to be able to apply that reasoning down to something even more specific. And (3) you also want to be able to generalize up to the more abstract. Besides, how unfulfilling is it to engage in activities you don't understand? Why would you want to do that if you don't have to?

I then gave them a classic problem from math folklore, the Bear Problem:

"Huh?" said everyone. "How is this math?"

I assured them that there is a mathematical concept behind this, even though it sounds like simply a riddle or joke. There were several insightful and creative attempts at scientific answers involving the angle of the sun and black shadows, and the 3 directions representing 3 colors being mixed. We were looking, though, for a mathematical solution.

I told them that I would give them a solution strategy by showing them James Tanton's Mathematician's Salute. I don't want to tell you much about the Mathematician's Salute because it will ruin your fun in figuring it out (search for the video online), but I will say that it physically demonstrates the "working backwards" strategy in a math problem: start at the end of the problem and you may discover a pathway to the solution.

Solution
Spoiler for readers: You can figure it out by the color of the bear. Where in the world do different colors of bears live? Brown bears? Black bears? White bears? Do any of these bears live in places where somehow the woman would return to where she started after she walked? Polar bears, you say? The North Pole? I didn't ask the students these questions—more leading than I'm comfortable with.

Doing that activity was enough of a hint for G to realize how the end of the Bear Problem could be used to get to the answer. With help from her math circle colleagues, they did figure out the solution together.

The solution to the Bear Problem led us back to our definitions. It made apparent a flaw in an attempted definition from last week. The kids refined their definition, and I congratulated them for once again thinking like mathematicians.

At one point in our mathematical conversations above, I held up a flat clock as an approximation of a sphere. Several students started looking for a more spherical object to use, but I stopped them.

Why am I stopping you from finding a ball-shaped object?

The students and I discussed the idea that in general, a goal in math circle is to think mathematically. Therefore, we'd like to move more toward abstraction. Therefore, a flat object is just fine.

At this point we were out of time. As we cleaned up, J showed me the numerous compass designs he had been creating at home all week to use as targets. "I really want a BB gun, and if I get one, I'd shoot it at these targets." So, while we strive toward the abstract in mathematics, what fun it can be to put it towards a practical application. Not only can we apply the abstract to the practical, we

Yes, I did ask the kids this sort of meta-question (a question about a question). I really want the kids to do the thinking, even the thinking about the thinking. I ask a lot of "why" questions. I wanted to be sure that I was facilitating deep thinking instead of providing solution strategies. I had taken a lot of notes at my first math circle training with Bob and Ellen Kaplan shortly beforehand. So I used my notes to guide my leading at first. I highlighted all of the quotes in my notebook and typed them up into a "Bob and Ellen cheat sheet." I studied these quotes every week before class. (This cheat sheet eventually became, with the collaboration of other MC leaders, the article "How to Become Invisible.")

can study the practical to draw out the abstract. Ever since N visited our math circle last week, his mom reported that he has been seeing circles, diameters, and radii in everything. He told her, "My mind will never be the same again."

WEEK 4

As had become our custom, math circle began with some definition work. Definitions had not been in my original "plans" for the course, but because I had been paying close attention to students' reactions to the content, I knew that this topic was very compelling to them.

What is a plane?
The whole group arrived at a consensus that a plane can be defined as a "never-ending flat surface without thickness."

"Oh no," groaned Z, "Now she's going to ask us the definition of the word flat!" I responded with—you guessed it—a question:

But aren't you asking yourselves what flat means?
The kids had to agree that yes, they were. Questioning and doubting were becoming automatic. (I was thrilled, of course.) Another thing that thrilled me was M's objection when I proposed that we use the solution to the Bear

Problem to explore the definition of flat. (True spoiler here; you've been warned.)

"But the earth doesn't really have enough curvature for a 2-mile walk to put you back where you started. I tried it when we walked our dog," he said, curiously. What we're doing in math circle, they're clearly taking home with them.

It was also clear that they were really enjoying the challenge of Euclidian constructions, so I decided to give them some handouts. (I had never done so before in a math circle, and still rarely do.) The kids tried a few challenges from a list of figures that can be constructed with a compass and straightedge. The challenge of copying a line segment was met with quick success by all, but then hands slowed and faces frowned at the rest. Time for a history break.

Alexandre Borovik points out that "In mathematics, what you did not tell is as important as what you did tell to learners. In a math circle, you only scratch the surface of mathematics. Where do you stop? What is cut-off point? … Why? For lack of time? Or because you did not want to touch something that is perhaps still beyond understanding of kids and can only confuse them? Teaching is one continuous act of decision making: what should students do next?" I sometimes find it hard to find a good balance between offering direction and following the lead of the students. Awareness of this tug-in-two-directions is a good first step.

While the kids continued their attempts, I recounted some anecdotes involving:

- Euclid;
- Euclid's Elements;
- and Abraham Lincoln's study of Euclid's Elements.

The kids had fun coming up with expressions of confusion in response to some Euclidian definitions.

EUCLID: "A *point* is that which has no part … A *line* is breadthless length … A *straight line* is a line which lies evenly with the points on itself … A *plane surface* is a surface which lies evenly with the straight lines on itself."

"Come again?" offered J. "Huh?" said someone else. "What the…?" from another.

Then their worked diverged. No longer was everyone working on the same problem.

- I hinted that some Euclidian constructions can be created with the Flower of Life. This hint was enough to get most of the kids going.
- M was determined to use his tools to find the center of a circle since "they have a tool at Home Depot that can do that."
- G helped me counter discouragement with the reminder that she had constructed a square two weeks ago.
- Some kids asked me to check whether their various attempts worked, while others did not want to see or hear what their colleagues were doing.
- To balance the pleasure of the individual challenge with our goal of collegiality, we constructed one figure collaboratively.

Then several students demanded,

We need a hint on how to create a perpendicular bisector of a line segment! I asked all who wanted a hint to come to the chalkboard on the other side of the room to consult privately. Five followed me, while two stayed in their seats. I showed them what I thought was only the first step. "She accidentally showed us how to do it!" R called unhappily to the two across the room. It turns out that I had shown them more than the first step. That teachering instinct of mine had come up involuntarily again (oops!). I had inadvertently stolen the joy of discovering for themselves. I returned it to them by giving them a new challenge:

Can you do the same thing in fewer steps?
As they continued to work, I talked.

- I told them that some view geometry as not just the basis of language, but as language in itself.
- I read aloud some descriptions of geometric symbolism in the field of Sacred Geometry.
- Then we talked about extraterrestrials.

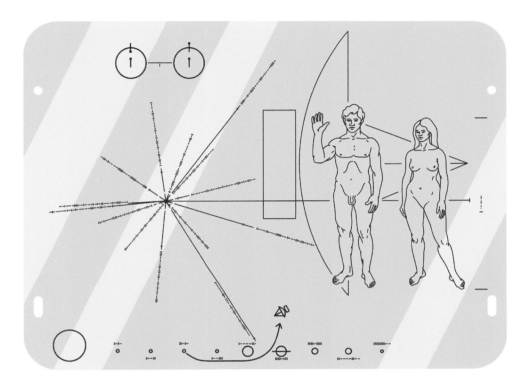

I showed them a (more modest) version of the message sent into space on Pioneer 10. This message uses pictures and math as language. The kids were able to figure out the meaning of the planetary order, but needed explanations for the rest of the message. They debated the existence of extraterrestrials for a moment before returning to their constructions. Then we ran out of time.

As people left the room, A and G ran up to Z, who had successfully constructed a perpendicular bisector in a very concise manner. A, who had been very quiet during the circle, demanded in an admiring tone, "Show me how to do it!"

"*Don't* show me," ordered R. Z led A and G into the corner to show them privately.

"I can hear you!" protested R, who exited, notebook in hand, still determined to figure this out for herself.

WEEK 5

We began the next session with a role-playing game invented by Maria Droujkova:

> Role playing in math drastically increases conceptual understanding because of the student ownership of mathematical material it requires. See Rodi's chapters on The Dark Bridge Problem and on Logic for more.

"Well, we don't want to waste erasers here in Alexandria, so *this* is the best method," argued R (playing the role of Euclid). She had just demonstrated how to perpendicularly bisect a line with only a ruler, a straightedge, and a piece of chalk. She faced off in debate against characters who might have different perspectives on this task: an Engineer (M, who spent the time before class theoretically improving upon the compasses we use); Artist (our visitor P); Computer Programmer (A); and Origami Maker (G). Since we didn't have enough assigned roles for every child, and, moreover, since the various perspectives weren't obvious, the kids immediately helped each other and played every role. They enjoyed imagining how the different characters would tackle the task.

"Code it!" said Computer Programmer A. "Measure it with a tape measure!" said Engineer M. "Just look!" And so on. Origami Maker was the trickiest perspective to fathom. Then I asked if they could think of someone who might have a method different from all of these. "Fashion Designer," was the reply. The kids explained that such a person would undertake a hybrid method involving eyeballing, measuring, and folding. Interesting, surprising, and true. Surprising to everyone because such a sophisticated approach is needed by the person/job stereotypically seen as the least intellectual.)

Then we returned to the Euclidian method. Last week several kids had vehemently urged me to allow constructions done using the eyeballing method. They had asked:

Now they understood why some people might prefer Euclid's method over just looking. "It's not perfect," said J about eyeballing. Then they contrasted the Euclidian method with measuring: which method would be more accurate in which situations? A then explained to P that the Ancient Greeks "didn't have a number system" that could effectively divide lengths in two, so it's a moot point historically.

We returned to our weekly challenge of attempting to define geometric terms.

What is a spiral?

Several children gave definitions which I wrote on the board. Then I read to them a formal, authoritative definition, which contained elements of each concept that was on the board. Smiles appeared on faces as M proclaimed proudly, "We really got it!"

Since P was visiting, we took a few minutes to show her some of the compass art we had been making at math circle and at home. As I was paging through my sketchbook, a few students noticed the Baravelle spirals I had made at home and said, "I want to make that!" (To make one, you construct a polygon, then use the midpoint of each side as the vertex of the next polygon to create an infinite series of inward spirals.)

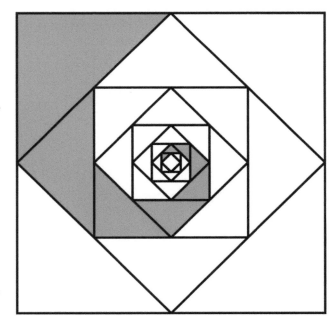

"Show us how to make that!" demanded G.

Do you all want to make Baravelle Spirals?

I asked the group, and heard a resounding, "yes!"

The group studied my Baravelle spirals. I explained how I had been attempting these spirals without a ruler, which is tricky since every line segment must be bisected. I told them that it doesn't matter to me how they did theirs. It was interesting to see the purists in the group engaging in the Euclidian method:

These students had come up with their own accessible mystery, and set to work. Others attempted their own challenges; some used rulers, and yet others eyeballed.

Our discussion returned to the pros and cons of each method. I mentioned how artists for ages have attempted to draw a perfect circle freehand. I also told of how I had filled years of boredom in school attempting this, unsuccessfully, on my own—and of how I spent years combing the beach for the perfectly circular stone. I did find one once, I thought, but was always afraid to check it with a compass in case it wasn't perfect. The kids begged me to bring the rock to Math Circle next week. I said I would look for it in my basement, as long as they promised to lie to me and say it's perfect even if it isn't.

As they worked, I told of the life and contributions of the ancient Greek mathematician, Hypatia. Then time was up. Everyone seemed excited to continue working on their spirals at home. And I was off to my basement to look for that rock.

WEEK 6

The final week.

"How many points should our polygon have?" R asked G. They were playing with Geogebra, a dynamic geometry software, before today's math circle began.

As I was setting up, I had mentioned to them that I hadn't yet found an obvious way to insert a diameter into a circle on this geometry program that I was trying out. Having never seen the program before, they got right to work on their own project: constructing a Baravelle Spiral in GeoGebra.

"I can move my point! This is so awesome!" said R.

"This could go on forever," observed G as they repeatedly enlarged the spiral to continue the pattern inwards. "This is better than anything on paper!"

"Way better," agreed R. The students were appreciating the beauty in infinity.

"We have to decide when to stop, though," continued G. I asked,

Can patterns get infinitely smaller?

Z walked in as we were discussing infinity. She looked at their work. "We're doing it freehand, without a compass!" announced both G and R from their spot in front of the computer. It took me a few minutes (and some explanation on their part) to realize that they were using the term "freehand" ironically, using a bit of Euclidian humor to say that this program delivered the goods in an immediate and precise manner.

"I think we ought to get started with our Math Circle," I told them, even though not everyone had arrived.

"But this *is* math circle right now," protested Z. I agreed, but explained that since today was this group's last math circle for a few months, I did want to spend some time showing them something else too. With a stricken look on her face, Z replied dramatically, "It's our last one? My poor, poor aching heart!"

J and M arrived as I was showing pictures of various compass designs and spirals: Baravelle Spiral quilts; nautilus shells; and mandalas from 7 different cultural traditions. As we were looking and discussing, most of the kids were already creating their own compass designs on paper. I told them the story of the artist Giotto's perfect circle and the Pope Boniface VIII's reaction to it, and we also spent some time defining the term "polygon." As I finished showing the mandala pictures, Z asked,

Do we get to make our own mandalas?

"Yes," I replied, and I showed them an unfinished mandala that I had created. They oohed and aahed until I told them that I couldn't finish coloring it symmetrically because the coloring step had revealed a major flaw in the design.

I asked,

Can you find the flaw and suggest corrective actions?

No one could, but everyone had their own project in mind.

- G wanted to create a mandala similar to mine, so she studied it and brainstormed how to create a dodecagram inscribed in a circle using the Euclidian convention.
- M and J were each creating BB gun targets. J's target was a Baravelle Spiral compass construction using a ruler (versus a straightedge) to quickly find midpoints, so he was coloring his intricate design before most of the kids had gotten a handle on their plans.
- Z was struggling with where to begin and requested, "I need a challenge!" M offered generously:

"Want to copy mine?"

She didn't, but accepted the same challenge as G, the 12-pointed-star mandala. I told of how that pattern had come to me in the middle of the night, after I had been brainstorming mandala designs for 2 weeks. I asked them:

Do you think "Eureka moments" are fact or fallacy?

People weren't sure. I told of the circumstances of Archimedes supposedly coining the word Eureka. I played an audio file of John Philip Sousa's telling of how his composition "Stars and Stripes Forever" had come to him in a dream. I also shared some related anecdotes about Einstein and Darwin. I concluded with an excerpt from an article about science writer Steven Johnson's book *Where Good Ideas Come From*, positing that the Eureka moment isn't a flash of insight, but what Johnson calls a "slow hunch." The kids agreed with this new thinking.

And that was the end of this course.

A math circle is, by definition, a conversation about math. And what a great conversation these six weeks turned out to be. If I am a journeyman in the math circle guild, for the kids this compass-art conversation has been a mini-apprenticeship in mathematical thinking.

EPILOGUE

It is now nearly six years later, and I still haven't found that rock, but I am still searching.

I'm also still trying to find that just-right accessible mystery, and voice those just-right comments and questions—those that will stimulate interest, discovery, and of course, more questions.

Am I saying here that every time you put out an interesting topic without a specific question that an accessible mystery will evolve? No, I haven't always had that experience. In fact, I've sometimes had the opposite experience—that highly-recommended topics fell flat despite all my presentations.

For instance, in another math circle, the topic of Leonhard Euler and his Circuits rose up like a phoenix from the ashes of the less-than-captivating topic (for my students) of Polydrons. Polydrons are a manipulative used for geometry. I had hoped to use them to delve into the platonic solids. In this course, before the students focused all their attention on Euler, they had come up with an almost-accessible mystery, "Can an elephant be represented geometrically with Polydrons?" Sadly, it wasn't enough to hold anyone's interest for more than 15 minutes. Several years later, I used Polydrons in a course on Escher and Tessellations, and the kids couldn't get enough of them. Fortunately, we can go with the flow in math circles.

Other topics did then emerge, though, as long as my attention was finely tuned to the students' questions, comments, and interests. (Keeping my attention finely tuned to the student's questions, comments, and interests is very hard. I do work on my own attention outside of class, and yet constantly struggle with letting go of my own agendas.) My conjecture here is simply that if you trust in the human architectural instinct, it will shine through if you let it.

On Mathematical Thinking and Math Circles: A Dialogue

Rodi and Rachel age 14

"This is really interesting," said Rachel to me during a writing session for this book. Both she and I were looking at some of our personal correspondences about people's math experiences. "This guy is telling me that something he hated about school math is how teachers always said that you have to learn math because you will need it in real life. But it turns out that in real life, he didn't end up needing any math beyond basic arithmetic." My conversation with Rachel then went something like this:

RODI: I wonder what he means by "math." I wonder: if he had been engaging in mathematical thinking, as opposed to memorizing algorithms, would he have benefitted later in life? Isn't that why Lincoln read Euclid's Elements as his bedtime reading? Not for the math, but to sharpen his thinking?

RACHEL: Well, that thinking is just analytical thinking, which applies to every subject, not just math.

RODI: Hmmm…Could there be something about the arena of mathematics, versus other subjects, that lends itself well to this type of thinking?

RACHEL: I don't know.

RODI: Or, could it be that when people use what you're calling analytical thinking (what I'm calling mathematical thinking) in some other discipline, that they're actually applying specifically mathematical thinking to that subject?

My discussion with Rachel petered out after another few minutes, but it left me wondering: is this type of thinking specifically mathematical? And, before even going there, what exactly is this type of thinking?

My immediate thoughts about what we're trying to do in math circle:

- let the kids invent math for themselves without telling them how to do it;
- help kids see what mathematics *really* is (to me, it's the creative pursuit of the underlying structure of things);
- shake loose their inaccurate assumptions about what math is;
- convey that there are deep philosophical questions involved in studying mathematics (I hesitate to even use the word philosophy here, in fear it will turn people off, but bear with me; I have high hopes you'll find this interesting);
- and set the stage for inquiry by asking interesting questions and letting kids collaboratively explore them.

According to one parent from my math circle: "In math circles, kids think hard about problems that may appear simple at first, but soon reveal themselves to be far more complex (and beautiful). By working collaboratively to come up with solutions to the problem, they learn not to give up (persistence pays off), and that there is more than one way to approach a problem (some more elegant than others). The biggest benefit of the math circle is that kids are empowered to think for themselves. They gain confidence in their ability to think critically and not simply rely on someone else's assertion of what is true." In other words, the math exploration done in math circle changed the child's thinking.

MATH RENAISSANCE

Says another, "The young people are designing learning for real life and real people." Another commented that once her seven-year-old son was in a math circle, "I noticed him wondering about the world around him in terms of mathematical ideas. He was asking questions in a way that clearly indicated to me that he was applying ideas he had explored at math circle to other things in his world."

Hmmm…these quotes make it sound like the pedagogy of math circles is generally applicable to a variety of topics, as Rachel posited. You might be able to facilitate these thinking skills in a Spanish class, or a history class, or a science class, etc. Well, maybe not Spanish—in that class you might have to rely on someone else's assertion of what is true. Wait a minute, there may be a problem extending this concept to history or science too: one goal of math circle is for you (the participant) to discover math on your own. This is possible; I've seen it done again and again. Can you discover history or science on your own? Maybe science, with enough time and access to living and dead creatures, fire, dangerous chemicals, and other fun stuff. In math we just need pencils. Could this just be a question of logistics, then? Hmmmm again…

I re-opened the conversation with Rachel.

RODI: Listen to what this one parent says: "Math circle was madly successful in launching questions in kids, questions that kept them wondering, working, trying to pose better questions and sorting out possible/impossible answers all week long." Says another, "I liked the focus on problem solving and the use of open questions and answers leading the kids to a conclusion. I also liked the incorporation

I'm going to fess up right now to the fact that I like to think that I'm not leading the kids to a conclusion. But I might be, actually probably am, some of the time. You also should know that I am not attached to the idea that Conclusion/Answer/Resolution is an important goal here, in our particular math circle. In fact, sometimes I even prefer that our 6-week sessions end without finding an answer to the question we asked on the first day. There's nothing like pondering math questions for years. I have a few 11-year-olds who pondered The Unicorn Problem for four years, and some 8-year-olds who worked on a logic question at ages 6 and 7 (that's almost 20% of their lives). And in both cases they liked the pondering.

of the history of math and biographies of mathematicians. I don't think I ever got that kind of thing growing up! I think it's helpful for the kids to think about the way in which real people struggled/struggle over problems and in many cases had to innovate in order to find adequate solutions!"

From these new quotes, it looks like we're fostering skills such as asking open questions, posing better questions, and coping with struggle over problems. I posit that we're moving toward the idea that the thinking skills up for debate here are more specific to mathematics. I think that any of the skills listed here in singularity can apply to other subjects, but taken as a whole, might be a skill set best developed through mathematics.

RACHEL: You just made a leap in logic though! You shared with me a bunch of quotes from parents about how their kids learned certain skills through math circles, and then you jump to the conclusion that because these kids learned these skills through math, they can't learn them through other subjects. That makes no sense!

I think you can learn these skills through English class too. In English, you ask open questions about things like symbolism, themes, and author's intent. You talk about things with other people who read the book, and then you refine your questions and ask better ones. Just like in math, it may take years to figure something out. You might never figure something out! With things like symbolism, there are many different ways you can interpret literary symbols, analogous to the many ways you can solve a math problem. In math and in English, your answers to questions might change over time as you get more exposed to the subject and as you mature. So, yeah, I think you need to think the same way for both subjects.

However, I do get a certain satisfaction from solving a proof or really hard problem in geometry that I don't get in English. But, I still think that you can develop the same type of thinking through English class.

RODI: So for now, can we say that we agree to disagree?

RACHEL: So you disagree?!!! I don't understand.

RODI: Okay, so maybe I haven't proven my point yet.

RACHEL: I don't see one skill here that you can't do in English.

RODI: Do you accept, then, that math is (among many things) the creative pursuit of the underlying structure of things?

RACHEL: Isn't that what English is too?

RODI: I don't know. Do you agree that people have an innate desire to uncover the underlying structure of things?

RACHEL: Yes, but I also think that English is about uncovering the deep truths in life and universal human emotions. Every book, if you get down to it, is really just about universal truths and emotions, and the things that we all go through, and that's what makes them good.

RODI: Well, I always hated English class.

RACHEL: Well, I don't like math class. Come to think of it, I don't really like English class either.

RODI: Well, can we at least agree that mathematics is universal?

RACHEL: And English is universal.

RODI: I don't know enough about English to accept that.

RACHEL: You don't know?!!! But it's obvious. English is the most universal subject in the world.

RODI: I think that math is the most universal subject in the world.

RACHEL: How is math the most universal?

RODI: When people sent those signals into space, they didn't communicate in English.

RACHEL: When I say English, I mean language. We're not talking language barriers here. And they didn't send math into space either, they sent pictures. Even if they sent math into space, we don't even know that there's extraterrestrial life out there.

The themes of English are the most universal way to uncover the deep truths about life as we know it. People always say that math is a universal language, but they're just talking about language barriers. They're saying that math is the same in every country. That doesn't prove anything.

Here, Rachel was pointing out a big hole in my proof. What was I trying to prove? That mathematical thinking is a specific type of thinking particular to mathematics that might be applied to other subjects. What was the hole in my proof? We hadn't talked about answers before. I didn't even mean "answers" exactly. I was using language loosely—a dangerous thing in mathematics. I should have clarified. At this point in the argument, I had forgotten about one thing we did talk about before: Rachel's earlier acknowledgement that you get something from mathematical proofs that you don't get in English. (By mentioning this, she gave me a chance to better define what I meant by mathematical thinking, to define it in a way that might support my proof. I missed that chance.) As a lawmaker, Lincoln used Elements as "a course of rigid mental discipline with the intent to improve his faculties, especially his powers of logic and language" (McClarey, Donald). He needed to know, as a lawmaker, what specifically is meant by a word he encountered frequently in his reading: demonstrate. He needed to know how to demonstrate the certainty of proof. In my argument, when I defined mathematics as an attempt to uncover underlying structure, I should have added the phrase "with certainty." Oops! Proof is a defining characteristic of mathematics. So I lost that argument! In our debate, each of us was really attempting a verbal proof. This debate sounds like many conversations in math circle. Of course, Rachel would probably say they remind her of English class. Oh well.

RODI: I do agree that English is an attempt to uncover universal themes, but not that it uncovers universal answers. Math might do that.

RACHEL: You don't think that English uncovers universal answers?!

RODI: Well, maybe I'm talking about the physical world only.

RACHEL: And I'm talking about the mental and emotional world. I don't think we even talked about answers before.

RODI: Hmmm...I wonder if neuroscience and physics might someday unite the physical and emotional worlds? Neuroscience is attempting to connect our actions and physiology to our thoughts and emotions. Physics, like religion, is trying to answer deep questions about why are we here and where did we come from?

RACHEL: Ummm?

RODI: Anyway, we do agree that math is universal. I'm not saying that English isn't universal, I'm just not well-versed enough in it to make such a declaration.

If you can tease out the bigger pedagogical picture here, you might have a better idea now of what I mean by mathematical thinking. What do you think? Are we engaging in a cognitive task that is specifically mathematical here? I don't know. Consider the mission statement of the Talking Stick Math Circle:

- to expose children to the richness of mathematics content
- to help children realize that mathematics isn't defined by arithmetic or performance
- to allow children to experience the creativity of mathematics
- to explore mathematics in a collaborative group, vs. competitive group or individual pursuit
- to help children find or increase enjoyment in mathematics
- to foster conceptual understanding of mathematical topics
- to inform parents about mathematics pedagogy so that they can increase their children's and families' math enjoyment and success in general
- to do so non-coercively

I haven't convinced Rachel yet. I haven't closed the gaps in my "proof." We're just playing with words here, not engaging in formal proof-making. With this essay, we want you to feel like you're in a math-circle conversation. But have I convinced you? I realize that this list, of course, doesn't really define mathematical thinking. I'd be negligent in attempting to prove something to you without defining my terms well. I need to do that. So also consider this

Stanford mathematician Keith Devlin describes mathematical thinking this way: "…a valuable mental ability—a powerful way of thinking that our ancestors have developed over three thousand years. Mathematical thinking is not the same as doing mathematics—at least not as mathematics is typically presented in our school system. School math typically focuses on learning procedures to solve highly stereotyped problems. Professional mathematicians think a certain way to solve real problems, problems that can arise from the everyday world, or from science, or from within mathematics itself. The key to success in school math is to learn to think inside-the-box. In contrast, a key feature of mathematical thinking is thinking outside-the-box—a valuable ability in today's world."

list of a few of the things that I think define mathematical thinking:

- asking questions
- refining questions
- identifying assumptions
- positing conjectures
- taking risks (if you're clueless, start with a crazy/wrong/impossible conjecture and use its wrongness to orient your thinking)
- proving and disproving assumptions and conjectures
- making generalizations
- moving from the specific to the abstract, and to apply formal reasoning to abstraction
- breaking problems down into bite-size pieces
- using analogous familiar problems/ approaches in unfamiliar scenarios

So, what do you think? Have I demonstrated that "mathematical thinking" just a math thing? Or do you think it's an everything thing? Or a math thing that can be applied to many other things?

We're not purporting to know all the answers here. In math circle—at least in this one—we're not in the business of providing answers. We're just asking the questions—the questions that might get you thinking that maybe there's another way. At this point, in fact, I think Rachel has me leaning her way. What's important, though, is that we hope we've left you wondering. If you are, then you're experiencing math-circle thinking.

Changing the Way We Teach Math

Rachel age 14

How do students learn best? Do tests measure how much a student learned? Do they measure how intelligent a student is? These are all big questions in the field of education. This book tries to answer these and more. Here we discuss what it means to learn, the best way to teach math, the problems with our education system, activities, and more. We talk about the value of inquiry-based learning. I wanted to tell you a personal story about this type of learning. Actually, it's my story. So here goes: the condensed memoir of a teenager.

I've experienced both homeschooling and public school, and I've noticed that the educational methods are very different in each setting. When I was homeschooled, I just learned things I was interested in. I read many, many books. I never took a formal English class, with grammar, literary tools, and essay writing. I took few classes, and was mainly content to learn on my own. If I had a book to read and a computer to do science research, I was set. I did elaborate science fair projects, where I researched online to understand the cutting edge research in a field so that I could form my research questions.

I learned math in many different ways. I always loved to shop. I shopped at the food co-op in my neighborhood ever since I was two years old. By the time I was three I learned how to count; "Put 5 mushrooms in the bag," my mom would say. By the time I was five or six I could use the scales to weigh and price the produce. "Get $2.00 worth of apples," my mom would instruct. I would have to see how many apples would equal $2.00. That's how I learned arithmetic.

I learned decimals and percentages from my science fair project, "Maximizing Flux in Forward Osmosis." I had to measure different concentrations of solutions and I found a need for decimals and percentages while doing this, so I learned how to use them. In another of my science fair projects, "Do Wind Turbines Stir Up Air Currents and Therefore Affect Erosion?" I had a need to describe central tendencies so I learned average, median, mode, and standard deviation. In fact, I learned most of my math through science fairs; whenever I needed to know new math skill for my projects, I would learn them. I also participated in and loved math circles (cooperative math workshops). I loved learning about compass art, Euclidean constructions, proofs, and mathematical paradoxes such as Bertrand's paradox.

People ask me all the time: "How did you learn anything?!" Well, research has shown that without coercive education, students will have internal motivation to learn. Although you might think that a student who wasn't forced to learn might just watch TV all day, that's simply not true. Children want to keep up with the standards of modern life; everyone wants to learn to read, for example. Children have natural curiosity. When placed in an environment that encourages creativity, self-expression, and inquiry, a child's curiosity will soar!

I have known kids who have gone to school, and once they started homeschooling, just wanted to recover by watching TV all of the time. This happens because school is exhausting! It is so competitive, the learning is so rushed, the curriculum is jam-packed, and once a kid has been to school for years they are going to be exhausted from schooling.

Also, sometimes subjects aren't taught in an engaging way in school. For example, one of my math classes was dull. We just did math problems, and learned algorithms to be able to do more math problems. We didn't have any challenging and engaging problems, no projects, and no big questions. I can tell that the whole class was pretty bored because half of them talked and half of them slept. I wasn't totally bored because I had done years of math circles before the class. I already learned about the beauty and intricacy of math, and I love it. It's not that I'm a natural math genius or anything, but I've found the beauty and mystery in it. I know that math can be lots of fun, but the other kids in my class didn't.

Teachers tell me that one problem in public schools is the curriculum. The teachers are given a very fast paced curriculum, interspersed with many tests. The subject matter covers a wide range of subjects, but none very deeply. The teachers are required to teach following the curriculum, even if they don't like it. For example, one of my math teachers told me, "We can't do any projects because we have to cover the whole textbook."

One contentious discussion in education is how and why the United States lags behind other countries in terms of test scores. The students here don't score as high, especially in science and math. Math education in particular can be a touchy subject. There are many different ideas in the United States about how our children should be taught math. Two such competing ideologies are "Reform Mathematics" and "Traditional Mathematics."

"Over the past 20 years educators have fought over the best way to teach numbers to kids. Advocates of traditional math tout the practice of algorithms and teacher-centered learning, whereas reform-math proponents focus on underlying concepts and student inquiry," writes Linda Baker in *Scientific American*.

Traditional and Reform philosophies are rather broad, but there are some more specific ideas about how to help students in the US score higher on math and science tests. One idea is to increase the hours of school here in the US. In some countries, kids go to school 6 days a week, or don't have a long summer break. This idea suggests that if the United States increased the number of school hours per year, the students might score higher on tests. This method may work, I don't know, but I think there's a better one out there.

Another approach is to change the curriculum itself, the way the students are being taught. The idea would be to let the students have more free time to explore and more opportunities to learn for themselves. Consider the example of Finland.

In Finland throughout the past 40 years there has been a *huge* wave of educational reform, resulting in almost no standardized testing, smaller class sizes, and more trust given to teachers to teach their students however they think is appropriate. This movement has worked; test scores have risen, and the whole world now looks to Finland as a world leader in education. Finland is now proposing a new plan: phasing out separate subjects and combining them in realistic ways. This will allow students to use the skills they learn in school in their real lives.

Finland is a wonderful success story, but sadly the US is not following in its footsteps. Although US schools using an inquiry-based approach have succeeded and prospered, they are not the norm. There is a wide variety of types of schools that use an inquiry-based approach, and some use it more than others. Some types of schools, such as democratic and free schools, integrate a lot of inquiry-based learning into their curricula. Others, such as some public, Quaker, charter, and Montessori schools, have integrated a little bit of inquiry-based learning into their curricula.

However, most schools in the US do not use any inquiry-based methods. Large class sizes are typical, substantial time is spent on standardized test prep, and teachers are given little trust to decide what is best for their students.

In math, Finland has certain methods that differ drastically from the United States' methods. Homework isn't as important in Finland as it is here. "I think meaningless math homework keeps girls in particular away from learning to love math. We have tried to develop a variety of teaching methods, including cooperative learning, problem-solving, concept attainment, role playing, and project-based learning," says Pasi Sahlberg, who wrote the book *Finnish Lessons: What can the World learn from Educational Change in Finland?* In Finland there isn't fear surrounding math, mutual respect is formed between the students and the teachers, and math is taught cooperatively.

Finland has removed all gender bias from math and the sciences, so boys and girls succeed at the same rates. Just how have they done this, you might ask? "To get girls more involved and interested in math, it is good practice to use different teaching methods. Girls want to work in pairs and groups because they can test their thoughts and share their ideas about how to solve the math problems. Both girls and boys like to do hands-on exercises and move out from their chair. The key is to give girls more differentiated instructions," says Maarit Rossi, a Finnish principal, math teacher, and the co-author of a new math program, *Paths to Math (Rossi as cited by Rubin)*. All Finnish textbooks have also been redesigned so they portray both boys and girls.

Math is among the subjects affected by Finland's move away from teaching by individual subject toward teaching by general topic. One example of this is kids enrolling in a course called "cafeteria services" where they would learn math, foreign languages, and the communication skills necessary to serve foreign customers. The Finnish justify their new education idea by saying that the real world isn't separated by subject, so why should school be.

Finland is doing a lot of things right. Their students love learning because it is made fun and interesting for them, and because of this, they succeed on tests. Hopefully the US and the rest of the world can soon follow in Finland's footsteps and start to reform schools so that students can really learn in a way that works.

There is no doubt that this would be a very hard change for us to make here in the US. The process of changing laws takes a very long time; In Finland the educational reform took 40 years. Also, in the US we don't control education centrally, but state by state. Furthermore, the people who make education laws in the US are politicians, not educators. These politicians often aren't aware of cutting edge research about how the children's brains work, and what the best way to teach is. They are also often motivated by other concerns like what their financial supporters want.

It might be easier if everyone could choose to send their kids to the schools of their choice, without regard to cost or location, but that is unrealistic. Private school is too expensive for most people, and although charter schools also exist in many places, it's not clear that they actually provide more innovative, superior educational experiences. Furthermore, this overall move towards school choice has the end result of diverting funds away from the already beleaguered public school system, which suffers even more.

Some public schools are really good, have innovative teachers, and successful students. Some private and charter schools aren't innovative, and produce students that aren't any more successful. In general, though, teachers in public schools don't have as much freedom to innovate because they have to follow stricter curriculums, have larger class sizes, and experience more overwork.

Everyone who is making a choice about how to educate their child should know all of the options. They should know how their child's brain works, and know the best way that children learn. Hopefully every kid will retain their natural curiosity no matter what school they're in.

I wanted to share my story with you because inquiry-based learning is hard to explain theoretically. It looks different for everyone. I wanted to give you a glimpse of what it might look like for one person, so that you could better understand the concept.

So, now that you (hopefully) have a better understanding of inquiry-based learning (and don't worry, we're going to be talking about it a *lot* more), what should you do? It's so easy to find flaws in our education system. All the time I find myself thinking "well, this needs to be fixed, and this, and this…" but whenever I try to come up with solutions I have a heck of a hard time. Just as in life, it's so much easier to find problems than solutions. So, when my editor told me that I had to give advice to you guys about solutions, I groaned in dismay. It took a lot more work than writing about why math education in this country has some room for improvement. But, honestly, how are we ever going to change anything if we can't formulate answers? There's no point in me babbling on and on about how the problems with our current situation without giving some suggestions about how to fix things. So, here goes.

To all the teachers and administrators out there: I know that you don't have a lot of freedom to choose what you teach and how you teach. But, when you do have a choice, please consider adding more inquiry-based learning to the classroom. It has so many benefits. Now, for the parents: hopefully from this chapter you have gleaned a better understanding about how an inquiry-based setting improves children's learning. I hope that with this knowledge you will be able to make better informed decisions about your child's education. For example, does their school teach in an inquiry-based way? Is there any way the school could change to become more inquiry-based with parent advocacy? Does your child have a strong opinion about how they would like to be educated? Does your child feel as if they are learning? If not, is it worth it to consider switching schools? These are all questions that can be asked when considering if your child is receiving the education that is right for them.

THE UNICORN PROBLEM

Rodi

"There is a unicorn dying at the end of a bridge. He has 17 minutes left to live, unless 4 people can join hands around him and recite a magical spell. There are 4 people on the other side of the bridge, but it is very dark, they have only one flashlight, and the bridge can only hold 2 people at a time. Ginny can cross in 1 minute, Ron in 2, Fred in 5, and Percy in 10. Can they save the unicorn?"

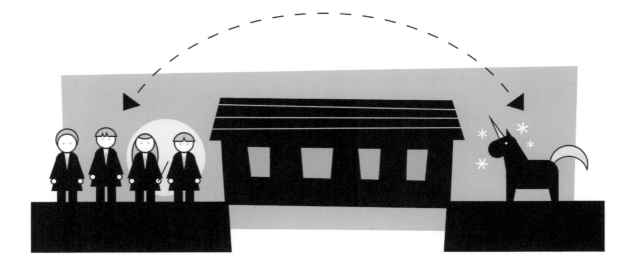

This is a really hard problem. Your first instinct is probably to add up the numbers in the problem, get 18, and say, "Can't be done." Then you'll really think about it, and realize that maybe something can be done. You'll struggle, and struggle, and struggle. You'll feel dumb. At least I did, as did my students.

But they latched onto the metaphor (story) and used it to push forward against mathematical failure.

What's the metaphor in the unicorn problem? These names are the same as some of the protagonists in the Harry Potter series. But why use stories and metaphors in math? Take a look at the following references:

- **Darmok and Jilad at Tanagra.** If you're a fan of *Star Trek: The Next Generation*, you'll know what this phrase implies. If not, know that it's a metaphor conveying an invitation for strangers to unite against a common enemy ("Darmok").
- **Festivus.** If you're a Seinfeld fan, you'll know it means something like "let's forget our differences, find common ground, and celebrate together" ("The Strike"). If you're not a fan, that's okay. (I'm not the biggest fan, but this line still invigorates me as a member of a certain generation.)
- **Ana seeking Elsa up the mountain.** If you have young children, you may recognize this as a dramatic point in the recent Disney movie Frozen. My take on this reference as metaphor is, "it's worth sacrificing anything to repair a rift and bring sisters together" (*Frozen*).
- **He who must not be named.** With this reference from Harry Potter, a whole generation of fans wants to unite against evil. (Rowling, J.K.)
- **Juliet on the balcony.** While the language of the scene is rife with metaphors, I contend that the scene itself is a metaphor: when I picture it, I feel the sensation of choosing a course, consequences be damned (Shakespeare, William).

Embedded in all of these metaphors are messages that people within a given community understand, that bring people together. Of course, even though the fandom of Harry Potter may extend to two million people, you may not get it, or these others. But you have your own cultural metaphors, right? How do you feel when they bob up to the surface? Metaphors and stories can be life rafts that motivate you to work harder to overcome adversity.

The adversity in this math circle was a collaborative attempt to solve the unicorn problem.

WHY IS
THIS QUESTION HARD?
- Deep mathematical thinking is required to answer the question. (Jump to the end of the chapter for a list of the major mathematical thinking skills involved.)
- You have to shatter an assumption to solve it. Our natural instinct is to assume that the fastest person, Ginny, crosses and carries the flashlight every time. Try that. It won't work. So, do the two slowest need to go over together first and get those slowpokes out of the way so that the speedy ones can sprint to the finish?
- Young students will probably not have developed the skills to successfully answer it yet, but hopefully will be sufficiently drawn in by the drama to demand to know how to do those things.

WHY
DO I LOVE THIS
PROBLEM?
Kids can't get enough of it—a horrible, mesmerizing struggle that forces students to question assumptions and think in new mathematical ways. In fact, I hadn't even planned to work this problem in this course. But many of the kids had come to a math circle demonstration class 2 months earlier in which we started to attack it, and demanded that we further pursue it now. That's the power of an accessible mystery.

THE STORY OF MY STUDENTS' STRUGGLE

My ages 6-7 math circle participants immediately began doing what mathematicians do: clarifying the question.

- *Can they throw the flashlight?*
- *Can they carry each other?*
- *How much does the unicorn weigh?*
- *Can they get across without the flashlight?*
- *Can one person go at a time?*
- *Did they bring any tools?*
- *Can the unicorn use magic?*
- *…*

They asked many questions beginning with those 2 magic words that spark mathematical discovery: "What if…?"

- *What if Ginny made return trips with the flashlight?*
- *What if Fred and Percy travelled together?*
- *…*

Some looked for loopholes in the rules (a skill that will come in handy for evaluating proofs) while others added up the numbers to see what is possible. While stymied with every attempt, enthusiasm grew. Many conjectures were produced. None led to a solution.

WEEK 2: MAKING SENSE OF THE UNDERLYING MATHEMATICS

What happens when a slower walker and a faster walker walk together? How long would it take Ginny (a 1-minute crosser) and Ron (2-minute crosser) to cross together?

Most kids had forgotten their conjectures about rates from the prior week. If it takes Ginny 1 minute and Ron 2, then together they must need 3 minutes to cross, right? The kids struggled with the idea of combined rates.

I relayed a historical anecdote in order to cleanse their mental palates. I wanted to then play Function Machines to work on the rate issue, but couldn't, as the ideas, excitement, and questions about this problem just kept coming.

We dramatized it with matchbox cars. All the kids then understood that the slower car determines the rate when the cars are "glued together."

Then we talked about attached people. A number of the kids did not transfer this concept of "attached" rates from cars to people and once again added the rates. Time for another dramatization. M stood in between P and V, then walked at her maximum but slow rate, while P and V held hands with her and walked at their faster maximum rate.

"Why do I have to be the slow one?" asked M.

"Because you are strong," I said. The need for strength became apparent when P and V pulled forward and M pulled back and they all finished at the slower rate. Then the whole group was able to make two generalizations:

- "Things can do less than their maximums but not more than their maximums.
- "Slow things hold back fast things."

Everyone agreed. I was curious whether they'd now be able to apply the same concepts to the unicorn rescuers. By that I mean would they be able to transfer a concept from a physical example to a purely mental construct?

They suggested that the two slowest walkers, Fred and Percy, cross together first. That would take 10 minutes. The faster one, Fred, would return with the flashlight. Now 15 minutes would have elapsed…

The kids were able to accurately predict and demonstrate the rate of crossing (math success!) but were sorely disappointed that the unicorn would die with this method. Kids started rolling around on the floor in defeat after a valiant effort. J remembered the safety provided by our ongoing Harry Potter life raft, and announced, "And the Nimbus 2000 whisked in and carried the unicorn to safety!" (The Nimbus 2000 is Harry's broom.)

With a few kids rolling around on the floor, another flying on an imaginary broomstick, and the rest still wanting to further test possible solutions to the unicorn story, this group needed to regroup/debrief. But first they needed relief. I quickly drew a broom on the board and said "Actually, the Nimbus 2000 is also a function machine." I told them how it worked and what sound it made and immediately everyone was sitting around the table, totally engaged.

With one minute remaining in session 2, one young mathematician jumped out of her seat and came up to the board to illustrate her solution idea.

Then parents arrived to pick up their children. One parent asked what progress we had made on the unicorn problem. I wanted to recap our progress for the kids too, so I told him (*very loudly*) that we now are convinced that a slower person's rate determines a faster person's rate when walking together, and we know that the unicorn dies when the 2 slowest people go first. One child said that next time we should try the fastest people first. J suggested doing it with "imaginary people." I asked the group if this problem would be easier to solve in their heads instead of by acting out. Since there is more than one way to enjoy math, each option had some support.

By this point, I was concerned that some of the students had given up. I reminded them that last week we had made two plans:

- to use M's suggestion that we see what happens when the fastest go first, and
- to use J's suggestion figure it out with our imaginations (instead of our bodies).

P was not at all interested in returning to this problem, and I suspected that A and N had lost interest too. The lifeboat of the Harry Potter metaphor was not close enough or strong enough to keep them afloat—or they had simply lost hope.

"I just learned that this exact same problem faced another group of students, who did solve it mathematically," I said.

"Now we know it can be done!" said M. Our question was no longer, *Can it be done?* but *How can it be done?* Fortunately, our group remembered how to combine rates, and was able to do the arithmetic.

Unfortunately, the unicorn still died when our characters Ginny and Ron crossed the bridge first and Ginny (the fastest) returned to give the flashlight to the slower walkers. I reminded the kids that this problem *can* be solved. The Harry Potter metaphor must have had some strength left because someone

What to do? I always fight the urge to rescue kids; rescuing them seems to make it all the harder to discard the crutches. However, compassion dictates that we not push our students past the breaking point. We need to stay attuned to them so we can help them stay grounded before they lose their focus and interest entirely. I wanted them to have hope, so I reframed the question.

proposed that we use our power to go back in time and give the unicorn another chance. We decided to make yet another mathematical attempt to save the unicorn next week.

WEEK 4 : TURNING AGAINST THE UNICORN

"Stories are constructs that can be reconstructed, but they are not free-floating. In other words, we co-create the world we live in … The stories that make sense of this world are part of this world … We transcend our world by being able to story it differently" (Loy).

Everyone still wanted to work on the unicorn problem. But a few kids in the class were actually hoping that the unicorn would die, and wanted to prove that saving it was impossible. (A single instance of reframing the question wasn't enough to keep everyone's interest. It was also necessary to switch allegiances. I wonder whether this new switching-allegiance/loyalty tactic arose from the power of the Harry Potter tie-in? Is that how Voldemort, I mean He Who Must Not Be Named, gained supporters?)

I told the anti-unicorn faction that in another world, a unicorn was living, and would only die if they could get to it within 17 minutes. Finally we were all doing the same problem, sort of. This time, with little coaching from me (other than helping them remember their failed calculations) and a lot of mental effort from themselves, the group collectively figured out how to combine the numbers to get 17. "The unicorn lived!" shouted M, as she and A started dancing around. "The unicorn died!" shouted O and P. J, N, and V stayed at the table, but I think that everyone was feeling satisfied that this math question posed many weeks ago was finally answered. (See end of the chapter for solutions.)

Would these kids have persevered for 4 weeks on this problem if we hadn't the life raft provided by the story/metaphor? Maybe. The non-Harry Potter version of this classic river-crossing problem has been passed down (under many names) through generations of math folklore. These problems can go beyond arithmetic and assumption-shattering into graph theory, combinatorics, and other areas of mathematics.

The original version of the problem (from math folklore) goes something like this:

There is a dark bridge that 4 people need to cross. One person can cross in 1 minute, another in 2, another in 5, and the last in 10. They are standing together with one flashlight, and the bridge can only hold 2 people at a time. Can it be done in 17 or fewer minutes? (Rote, Günter.)

It's fun to talk to students about the long history of river-crossing problems. "The wolf, the goat, and the cabbage" problem dates back to the 9th century: A man has to cross a stream in a boat that can hold himself and only one other object. He needs to transport a wolf (or a lion, or a jackal), a goat (or a sheep), and a cabbage (or bundle of hay, or pumpkin). He must be sure that when he is out in the boat the wolf does not eat the goat and the goat does not eat the cabbage (National Council of Teachers of Mathematics). Another of my favorites is the "missionaries and cannibals" problem: Three missionaries and three cannibals must cross a river using a boat which can carry at most two people, under the constraint that, for both banks, if there are missionaries present on the bank, they cannot be outnumbered by cannibals (if they were, the cannibals would eat the missionaries). The boat cannot cross the river by itself with no people on board (Pressman and Singmaster).

Definitely still a compelling problem. But would the repeated failure over 4 weeks keep a bunch of 7-year-olds hard at work on it? I don't know. But I have seen that narrative stories have the power to provide the support needed to endure struggle. And you probably noticed that a big assumption I'm making here is that struggle is not A Thing To Be Avoided. In math circles, leaders hope to foster a tolerance for struggle. Struggle, which is not the same thing as drudgery, is integral to mathematics. Obstacles are rampant. Exploring problems can be an emotional roller coaster.

If I'm right about the power of the narrative, the following quotes should illustrate the value and rewards of struggle.

- "A smooth sea never made a skilled sailor."
- Math-and-mindfulness educator Richard Brady explains that when his "students encounter obstacles, their first impulse is usually one of two extremes: they try to overcome them or give up. The approach of welcoming obstacles, sitting with them, and seeing what gifts of understanding they have to offer is foreign to my students—yet it is one that could serve them well in life. I ask myself how I can do a better job of modeling this way of relating to difficulties in the classroom. I realize I could begin by curbing my impulses to diagnose and suggest remedies for students' problems and learn how to just be with the students and their problems."
- Mother Teresa (supposedly) said it succinctly: "Life is a struggle; accept it" ("Quotes Falsely Attributed to Mother Teresa").
- "When you understood a thing, it gave back to you, could bring you energy though you believed you were expending energy to do it. The paint would fight you unless you understood it, just like the weather. Just like people" (Ruby, Ilie).

- "A religion was not a race. It was an idea, and ideas stood (or fell) because they were strong enough (or too weak) to withstand criticism, not because they were shielded from it. Strong ideas welcomed dissent" (Rushdie, Salman).
- "He that wrestles with us strengthens our nerves and sharpens our skill" (Burke, Edmund).
- "Our antagonist is our helper" (Rushdie).
- "What doesn't kill me makes me stronger" (Neitzsche, or Kanye West, or any number of people as cited by Grow).

EPILOGUE

Four years later, I revisited this problem in a math circle with some of the same kids, and few new ones. On the first day, we joined together on blankets under a tree in a garden in the spring. I announced that we are revisiting math circle history in this course. "Are we going to do the unicorn bridge problem? That's my favorite math problem of all time!" said M, who had done this problem in math circle nearly 4 years ago.

Immediately the kids got to work. The literary/dramatic/metaphoric connection held as the kids started to work the problem. But would this connection carry the kids, now older, through the trial of the difficult math involved?

They posited and tried numerous conjectures. Every conjecture failed. Determination was faltering. "C'mon, guys," said M encouragingly, "we solved this problem when we were like six!" I reminded the kids who had been in that math circle years ago that it took a larger collective of kids six weeks to solve this. Also, when they were younger, it took four weeks for the group to understand that if a slow and fast person walk together, the slow person determines the rate for the couple. Four years later, it took about 3 minutes for the students to come to that conclusion. This conversation gave them a bit more hope.

Another thing that happened much sooner this time around is the dropping of the assumption that the fastest person has to go first, and that the fastest person has to go multiple times. (Interesting, since my meta-assumption has always been that the younger you are, the looser your attachment to assumptions. Hmmm....)

Soon, this group was attacking this problem like a combinatorics problem—trying to exhaust every possibility of combinations/orders. But they didn't have a systematic way to list all the possible arrangements, so they got tired. They decided to do what mathematicians do and work on a different problem for a while.

The next week toward the end of class, the kids finally felt ready to revisit the problem. Some of the students were excited to work on it, but others had become discouraged, particularly those who had been there 4 years ago when the group solved it over six weeks. "Why could we solve it when we were 7 but we can't now that we're 11?" asked one frustrated student.

Again I reminded them that "for one thing, we spent much more time on it then. That was the only problem we worked on for weeks. For another thing, we had a much bigger group—more minds working together."

"I think there might be no solution," said another student, who had not been in that original group. Those in the original group were even starting to doubt their own memories. I assured them that there is a solution, and that it was within their collective grasp. This reassurance gave them the willingness to sally forth. They proceeded with 2 weeks of acting out the problem. The recurring argument over who had to play the dying unicorn definitely hindered progress. I kept a running record of their conjectures and results on the whiteboard.

Four weeks in, during the final session, L suggested that we look at every attempted solution so far and see if we could figure out what arrangements the group had not tried. So we looked at the whiteboard with our cumulative work. Unfortunately, over the weeks some of the work had worn off the board. Fortunately, I happened to have their work typed up on paper with a copy for everyone. I passed these out.

L went right to work on her own checking and rechecking different permutations. Everyone else focused on the board and worked together on re-trying old attempts (double checking the work so far) and coming up with new ways to try it. The boardwork, directed by the students of course, ended up looking a bit like a probability tree. The group finally had a systematic approach to the problem. Hope was rekindled.

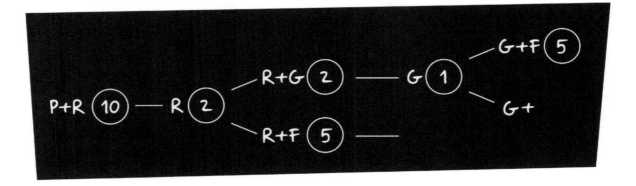

Hoping to name the group's global strategy, A asked, "So what are we doing here? Are we just trying all the combinations? How do we know if we've identified all the combinations?"

"There are hundreds of thousands of combinations," answered R.

"Not that many," replied someone else.

"A number that we do not know," countered R.

"What would you call that number?" I asked. We worked in this manner until we ran out of time. The students now had a systematic strategy for their combinatorics problem. It was a shame that we were out of time. Most of the students seemed happy to have their handouts and their strategy so that they could continue working on this problem as the mood arose at home.

And that ended our math circle for the year: an unsolved problem, but not an open problem, and people with the tools and curiosity to solve it.

It did seem that some of the magic was gone from this problem compared to our attempt 4 years ago. Was this because some had done it before? Or was it because by age 11 the draw of the narrative is less compelling? Maybe. Or was it that phenomena many parents and teachers mention: that students (especially girls) seem to become less confident or interested in math during middle school? Even if the magic had somewhat diminished, it was not a bad way to end; no resolution, but a realistic and hopeful math problem to ponder.

What are the implications of the kids having solved this problem when they were younger but not now?

Should I have given hints? They were pretty close at the end, and the only thing stopping them was time. What do you think?

Despite what I know about humans as storytelling animals, for months I believed that the kids, once older, had just outgrown this Dying Unicorn story/ metaphor. But now I'm not so sure. Consider:

- "The world is made of stories, not of atoms" (poet Muriel Rukeyser).
- "Language is basically for telling stories. . . . A gathering of modern postindustrial Westerners around the family table, exchanging anecdotes and accounts of recent events, does not look much different from a similar gathering in a Stone Age setting. Talk flows freely, almost entirely in the narrative mode. Stories are told and disputed; and a collective version of recent events is gradually hammered out as the meal progresses. The narrative mode is basic, perhaps the basic product of language" (cognitive neuroscientist Merlin Donald).

- "It is with our stories that we make sense of the world. We do not experience a world and afterward make up stories to understand it. Stories teach us what is real, what is true, what is possible. They are not abstractions from life (though they can be that), they are necessary for our engagement with life" (zen teacher and writer David Loy).

Again I ask, What do you think? Rukeyser's realm is social commentary; Donald's is science; Loy's is spirituality. Am I being overly bold by extending these stories to mathematics? Is a narrative thread useful, indeed necessary, for full engagement in mathematics? Moreover, are we using stories even when exploring purely symbolic mathematics? Is anything, math or otherwise, purely symbolic? As long as I'm going out on a limb, let's ask this: is mathematics itself a form of storytelling? Is it, as philosopher Ken Taylor says, "simply fiction with certain constraints"? Hmmm…Now we're in a typical older-student math circle discussion. Welcome aboard.

SPOILER ALERT - SOLUTIONS Here's the solution my students came up with: First Ginny and Ron cross (2 minutes). Ginny returns (3). Fred and Percy cross (13), then Ron returns (15). Lastly, Ginny and Ron cross together (17). Another solution is that first Ginny and Ron cross (2 minutes) and Ron returns (4). Fred and Percy cross (14), then Ginny returns (15). Finally, Ginny and Ron cross together (17).

Boredom

Rachel age 14

Imagine: you are in math class. The room is stiflingly hot. Your teacher is droning on and on about factoring; you really couldn't care less. The kids next to you are passing notes, quietly murmuring. A stifled burst of laughter. Your eyes start to droop. Your desk is such a comfy pillow. You'll just close your eyes for a few minutes, and then you'll pay attention. Just a few minutes. . .

Many people want better math experiences for their kids, but only a few seek to change the system that has created all of these problems in math education. If you are reading this book, you are one of them. This book is all about creating better systems to improve mathematics education.

Changing the system is great and necessary, but it takes time. Right now, the reality is that not everyone is in great math classes. Sometimes people get bored in uninteresting math classes. This has happened to me. How can students make the best of the situation they're in if they can't change it? How can you fight the boredom? Here is advice you can pass on to your children and students.

Let's face it: everyone gets bored in math class sometimes. Sometimes this boredom reveals that there is a serious problem. Other times it's just an occasional thing and not to be worried about.

I was curious, so I asked lots of different people—students of all ages, adults in math-related fields, adults in non-math-related fields, teachers, and parents—what they do in math class when they are bored. Then I wondered: is this what they *should* be doing (if there is a should)?

I got interesting responses. Some people suggested trying to learn math in different ways, such as working ahead or thinking about the math topic in a different way. Some people suggested doing something else in class, such as reading, doing homework for other classes, inventing games, passing notes, or practicing different kinds of handwriting.

I asked all these people because I was stumped about what to do when I was bored in class. I thought that I had already tried every possible coping strategy. But when I asked for advice, I realized there were so many ideas I never thought of. I grouped the responses I got into different categories and included many of the quotes I received. All of them are ideas from different people; none of them are from me.

DOODLING

When faced with boredom, some people perfect their handwriting and drawing. "I spent years trying to draw the perfect circle (freehand). I also played with different kinds of writing: bubble letters, bubble letters with shadows, cursive bubble letters, cursive with shadows…all the while, listening to the teacher and processing the information. I wonder whether my 'doodling' actually allowed me to have deeper insight into the class material?"

Some people process information better when they doodle. Studies have shown that doodling repetitive things can actually help you think deeply about the task at hand. However, activities that require more cognitive attention aren't helpful, and actually are distracting and inhibit learning.

USING YOUR BODY AND MIND TO CHECK IN

"I would suggest trying to think of a way to teach this material to a young child. Forget how your teacher is doing it, how would *you* make it fun and playful? How could you make this into a game? Got any young children in your life? Try to teach it to them. Maybe they'll ask an interesting question that will make it fun."

A way to truly cement concepts in your mind is to teach them to someone else. It helps you understand things when you can explain them to someone else, especially to young children with whom you have to use accessible language. It's always fun to make things into games and fun activities!

When you are really bored, sometimes you can resort to very funny things, like "pulling a nose hair." Haha. On a serious note though, when you engage your physical body with an intentional action, your attention is sharpened.

Mindfulness can alleviate stress and help you live in the present moment. "I usually write poems and songs. Some people with artistic skills find cracks on the walls and imagine them as part of a picture, as my brother does. You may also do mental scanning of your body, noticing which muscles are tense, and trying to relax them—a little yoga here and there. Meditation teachers may suggest noticing all sounds."

Mindfulness, when practiced outside of class, can help with concentration and actually increase your attention span for when you're in class. However, practicing it in class is a way of checking out.

DAYDREAMING TO CHECK OUT

Some people doodled and did other non-math related things in class. (Just to be clear, I'm not advocating not paying attention in math class, I'm just sharing uncensored quotes from people I interviewed.)

- "I doodle as well…and practice my cursive writing."
- "Definitely a doodler too. Also wrote notes to friends, chewed gum focusing on not being caught rather than on the functions being taught. Go through collection of lip smackers. Smell and apply. Or look at my trolls fixing their hair into perfect points."
- "I was a note-passer extraordinaire. I think I should have had your book to read."
- "I would do homework from other classes, write notes with friends, or make to-do lists. The fact that I passed any math classes was a miracle."
- "My son invents games, not just computer games, but card and board games with interesting scoring systems whenever he is bored."

- "I see the teachers as something that I think of on the spot like giant marshmallows screaming something with gummy bear classmates or a lady screaming to robots. Or both robot gummy classmates."
- "I would imagine riding on a beam of light, observing what that would do to space and time."
- "Daydream."

How is giving voice to these "disruptive" comments productive? It's a way to acknowledge people's feelings. If we want to improve how math is taught, we have to acknowledge that some people associate it with boredom or suffering and simply tune out.

YOU MIGHT HAVE SERIOUS PROBLEMS WITH MATH CLASS

Different kids work in different ways and at different speeds, and it really hinders a kid's enjoyment and progress if their math teacher can't accommodate their type of learning. "Back in Vocational school, I would often race through any work teachers gave me as quickly as possible, so I'd have time left before the end of each class to work on *my* work. I was writing (and drawing) violent crime comedies. Bosses at various jobs did not appreciate this kind of behavior. They stupidly didn't grasp that their work *was* getting done, better and faster than anyone else would do it. The world is *full* of idiots, and many of them become bosses."

Hopefully teachers and bosses can accommodate kids who work in their own way. Not everyone works at the pace or in the way the teacher wants, but just because someone works differently does not mean that the result of their work is any less excellent. If kids were able to work at their own pace and in their own style, some kids would enjoy math a lot more.

The same person continues: "In 6th grade, however, I had a teacher who had a book that was self-paced that explored mathematics more conceptually—including many questions that drilled into my head that I couldn't divide by 0. It had 10 multiple choice questions per section with the answers available to check my own work. I worked on this once I finished my classwork or homework. While he wasn't one of my most memorable teachers for teaching me math, I do remember him for recognizing my interest in math and giving me a chance to advance my mathematical knowledge on my own. That was the only time that a teacher was able or willing to help me advance my abilities in math beyond the curriculum in the classroom."

It's lucky that this person had a teacher who recognized that students learn at different paces and ways and had the resources to do something about it.

WHAT TO DO IF MATH CLASS ISN'T FOR YOU

Some kids are bored in math class because the class is too easy for them. "What about asking the teacher for challenge problems that are about the topic on which your teacher is teaching? That is not what I did, but it is what I hope a student in one of my classes would do. I remember someone from last summer's Math Circle Institute at Notre Dame saying she puts a series of problems on a 'worksheet' in increasing difficulty so that students who breeze through the part that is 'required' have more to work on." A problem in math education is that some students are bored in math classes and aren't challenged appropriately. At many schools, like my school, there are more kids who would be appropriately educated in an honors geometry class, for example, than can be accommodated. Therefore, some children are bored in regular math classes.

It is really helpful for students when their teachers can give them additional/harder work to do, yet due to high workloads this is not always possible. If your teacher can't give you harder problems, someone suggested that students can "attempt to change the problem so that it is more challenging." It sounds fun to make up your own problems.

Another possibility is for students to do supplemental math activities outside of school, such as math circles. If, however, the student's school gives a lot of homework, they might not have time for any after school activities.

Sometimes students' situations at school are so bad that alternative educational methods are proposed. "And if I found myself bored day after day, or if I found I had no time outside of schoolwork to pursue making and playing and teaching, I would ask myself: is it worth my time to continue this path? Maybe taking regular high school classes is easy and convenient and it will help unlock other doors. But maybe the answer at the end is to find another way to get there (college? career?) from here…college classes? online classes? home school? internships?"

If students struggle with school, some people suggest alternative paths. These paths are feasible for some students, yet for many in public schools is the only option. If both parents work full time, then it is very hard to homeschool or cyber-school. Private schools are very expensive and are not an option for a lot of people. So, for one reason or another, for a lot of us public school is the only choice.

MY REACTION

As you can see, I received a *lot* of responses. When I read through them all I realized that I'm not alone: lots of people don't like school math! Before I read all of these responses, I knew in my mind that *of course* a lot of people get bored in math class.

When I read these, I felt better emotionally and my feelings felt justified. But I also felt sad. So many people have terrible experiences in math class! It's really a shame. So many people are bored from mind numbingly dull lessons. Sometimes this boredom is occasional and isn't really a problem, but sometimes the boredom reveals serious problems with math class. Not everything can be solved by doodling.

Even though I felt depressed when I realized this, it also makes me want to make a difference and *do* something about it. It's therapeutic and empowering to work and change the system. The system isn't likely to change while I'm still in school, so I wanted to compose a list of ideas for what kids in school can do *now* to make the best of the situation.

Still, even though it will take a while, I *do* want to change the system. I want to make a world where kids can learn any kind of math they want, where they can move their bodies while learning, and where there isn't such an emphasis on standardized testing! I know we all want a better world for our children, so let's do something about it.

Problems Math Teachers Face

Rachel age 14

For years, I've heard from kids and parents that they hate math, or that math was ruined for them, and I've wondered why. Why do some students struggle to succeed in math? Why do some people report that teachers, who try so hard to help children, sometimes come across as mean and boring? How can a nice person who means well get into this situation?

Well, I decided to delve deeper into this situation and interview a couple of people. I hoped that from talking to some teachers about the struggles they faced I would understand what parts of their situation they could change and what parts they couldn't.

I interviewed one of my high school algebra teachers, and she said that math was ruined for so many children because they think that arithmetic defines math.

"Arithmetic is just a tool; real math is so much more." Arithmetic is the type of math that is taught to young children in elementary school; it is their first encounter with math. They grow up thinking that arithmetic is the only type of math. When students struggle with arithmetic, they think they are bad at math. My teacher made the analogy that, "If there is a sculpture, arithmetic is the tools and chisel, and math is the sculpture. The mistake many people make is to think that arithmetic is the sculpture." Another math teacher at my school made the analogy that, "School math to real math is like spelling to English."

I interviewed a middle school math teacher, Todd Campanella, about his experience teaching in the public school system. He teaches math to children who have already failed a math course, so they have to retake it. He said, "At home, children hear their parents complaining about how they hated math when they were children. The kids might have been unsuccessful once, and thought that they were bad at math because their parents were, and think that they weren't born with the math gene. But they should try and not give up."

This is really true, and goes back to arithmetic being the first type of math being taught. Once the kids go to school, and have one unsuccessful math experience, they are overwhelmed. Because of belief in the math gene, they think they shouldn't try hard at math because they will never be good at it—and then they fail. It's all a cycle. The parents don't have a good math experience, so neither do their children, and neither do *their* children. This is one reason why math is ruined for so many people. It isn't the only reason, however.

"In public schools there is a lot of testing," Mr. Campanella continues. "Five years ago there was only one test per year for little kids, but now there are more for practice. Teachers are fired if students fail. The students in a class are all at totally different levels, and teachers have to teach all of them. Most students fall apart at long division, because it is multi-step. It's hard for them to remember how to do it." The fact that teachers' jobs are constantly on the line creates a level of tension in the classroom. This explains why sometimes teachers can come across as resentful when students ask questions. The teachers are just really stressed out about having all of the students prepared in time for the tests. Many educators, students, and parents believe that our educational system is too test-based.

"Teachers' jobs are on the line, and students don't pass," continues Mr. Campanella. "The students are bitter because they have a double period of math if they fail. But they don't really improve much mathematically. There is a push against memorizing facts, but facts are able to give you confidence and better grades. Lots of kids aren't confident. They see smart kids and are intimidated. But the smart kids have parents who work with them on schoolwork."

Teachers often talk about the impact of school budgets on their teaching. Budget cuts are disruptive. Also, budget inequality between districts hurts students because the infrastructure is designed for the districts with more money.

For example, the government has threatened to close the entire school district where I live a couple of times, because there just wasn't enough money to keep the buildings open. There aren't enough teachers, and in some cases there are sixty kids in a classroom. Some schools don't have nurses, counselors, or subs. For the 2013-2014 school year, all clubs, sports, and music programs were cut (Fund Philly Schools). And, in the past year, 5,000 high school students didn't earn enough credits in basic classes to graduate because of staffing shortages. Historically, our school district has had so many problems that fifteen years ago it was handed over to the state, which has controlled it ever since (Education Voters of PA). So, pretty tough situation, huh? Anyway, let's get back to the story.

Teachers also have a lot of responsibilities aside from teaching. "They have to do a lot of business work, there are also snow days, assemblies, and early dismissals," Mr. Campanella comments. All of this cuts into instructional time.

As I've mentioned before, a lot of the problems with math instruction start in elementary school. I interviewed an elementary school math teacher, Elaine Teitelbaum, about her experiences teaching in a poor inner-city public elementary school. She has mainly taught first grade.

"Everything depends on the teacher, but the teachers are stuck to the curriculum. If they don't stick to it, they get penalized. Some have more free range. It all depends on who the presenter of the material is: they have to be passionate and capture the audience."

Even though she mainly taught first grade, she also taught third grade and fourth grade a little bit.

"All we did was get them ready for tests," she reports about her experience with the older grades. "Students with more acumen for math seemed to work much harder. It would be more enjoyable if there were no restrictions for teachers. They can get penalized if they don't follow the rules. This is better in private schools. [In public schools], if they get in trouble, supervisors will come in. The Core Standards must be followed. It's a difficult position. The teacher is required by the district to teach in that manner. There are strict guidelines, they want consistency and equal opportunities for everyone. Everything is standardized. Yes, everyone should be taught some of the same things, but with more freedom."

In other words, teachers could do their jobs better with more trust placed in them. I was surprised talking to her because I didn't know teachers faced such harsh consequences if they strayed from the curriculum. It sounds like teachers could use more freedom.

National Education Association (NEA) President Lily Eskelsen García explains the devastating effects of the No Child Left Behind Law: "NCLB has corrupted what it means to teach and what it means to learn. Teachers have to teach in secret and hope they don't get into trouble for teaching to the Whole Child instead of teaching to the test" (Garcia as cited by Walker).

Because of the increase in testing, other "nonessential" classes are getting cut in schools: "Kids in urban areas are in most need of a well-rounded education and yet they are the ones who have had it stripped from their classrooms—and they don't have other avenues available to them that students in suburban communities have to at least partly supplant what is missing in schools," argues Richard Milner, director of Urban Education at the University of Pittsburgh (Milner as cited by Walker).

This is the same phenomena that Ms. Teitelbaum has witnessed. She told me that the awful home environments these kids are in break her heart and she just wants to help them, yet has no freedom to teach what she wants.[1]

"High-poverty schools across the nation have been forced to narrow the curriculum much more drastically than wealthier schools—with worse consequences for low-income students. While their more affluent peers may routinely visit museums or other cultural resources, many poor urban and rural students rely on their teachers to expose them to the kind of background knowledge that is essential to subject mastery," explains Tim Walker, an NEA author. Hearing this, I wonder what an administrator's perspective on standardization is?

Ms. Teitelbaum explains that another possible problem is that "the teacher might not be well versed in math." Why would this be a problem? In *Knowing and Teaching Elementary Mathematics*, author Liping Ma presents research supporting Ms. Teitelbaum's statement: "Having considered teachers' knowledge of school mathematics in depth, I suggest that to improve mathematics education for students, an important action that should be taken is improving the quality of their teachers' knowledge of school mathematics … It does not seem to be an accident that not one of a group of above average U.S. teachers displayed a profound understanding of elementary mathematics. In fact, the knowledge gap between the U.S. and Chinese teachers parallels the learning gap between U.S. and Chinese students revealed by other scholars." If we really want to improve mathematics education for students, then we need to improve mathematics education for teachers.

Dr. Ma writes that teachers don't have many opportunities to deeply learn and understand math:

"Teachers who do not acquire mathematical competence during schooling are unlikely to have another opportunity to acquire it … Most U.S. teacher preparation programs focus on how to teach mathematics rather than the mathematics itself. After teacher preparation, teachers are expected to know how and what they will teach and not to require further study."

It turns out that elementary school teachers teach all subjects, yet there aren't strict regulations about who can teach elementary school math. As we've seen, the most fatal mistakes in a child's math education happen in elementary school, when the child is first introduced to math. This is a time when it is crucial to have the most highly trained teachers. As Dr. Ma reveals, even though a lot of teachers are amazing, sadly, some aren't well-trained. Instead, they were taught badly in elementary school, which contributed to a lack of mathematical success. And since they felt they didn't have the math gene, they never excelled at math. When they were adults, and wanted their elementary school teaching degree, they still didn't have a full idea of what math really was, because they were just taught arithmetic and memorization. There aren't strict regulations about who can teach elementary school math, so they became math teachers without ever understanding what math really is. Then they teach children, who then have the same problems. It's all a cycle.

If all teachers received a thorough grounding in mathematics, as they do in China and Finland, and were trusted as professionals, as they are in Finland, then they could really have the freedom to do their jobs.

If you try to find one person, institution, or pedagogy to blame for all of these problems in schools, you can't. They are all too interrelated. It's a complex problem. Sure, we can list some of the issues: standardized testing, budget cuts and reduced funding, the belief in a "math gene," no trust in teachers, not enough days of instruction, and so on.

Furthermore, there are certainly more factors than those I've listed here. It would be reductionist to try to find an "easy fix." There really isn't one. It will take many years of education reform to eradicate all of the current problems with our education system—but it's not impossible. It's time to roll up our sleeves and get to work.

So, what can parents, teachers, administrators, and students do? I hope that for all teachers, reading these interviews has helped you realize that other teachers endure many of the same struggles you do. I hope that this helps you realize that you're not alone, and gives you hope for change, because like-minded people can band together to enact change.

To all the non-teachers out there, I hope that reading these interviews has helped you relate more to teachers and understand their challenges. I know that it's sometimes easy to harbor anger towards teachers that really should be directed elsewhere. It's not the fault of teachers that they have to follow national curricula with core requirements.

I'm hoping that an improved understanding of teachers' situations will infuse interactions with more compassion, and will help you make better, more informed decisions about your child's education. Hopefully you can identify what kinds of things are beyond the power of a teacher to change. With this information, you and your child can discuss whether they are receiving the education they need and whether that's going to change anytime soon.

[1] Even though NCLB is no longer in force, the testing requirements for math remain the same (Wong).

Logic, Puppets, and Kids Taking Over

Rodi

What happens when puppets do math? What happens when eight-year-olds operate the puppets who are doing the math? What happens when a puppet is the medium through which an eight-year-old channels her mathematical thinking?

This is the story of how Waggy the fox and Penelope the pig deepened children's understanding. No, wait. This is the story of how the children deepened their own understanding because they (1) immersed themselves in drama using puppets, and (2) took ownership of mathematical content through self-directed learning. That content was logic, a foundational principle of mathematics, in fact the basis of certainty in mathematics. We need it for proof.

ABDICATING POWER – THE BEGINNINGS ROOTED IN INQUIRY

"All puddings are nice.
This dish is a pudding.
No nice things are wholesome."

I put this little bit of math history on the board: one of Charles Dodgson's (aka Lewis Carroll) logic puzzles. The kids' curiosity lit up from just having this bizarre-sounding thing on the board:

- *What does that mean?*
- *What do they mean by pudding?…by dish?…by wholesome?*
- *Whose words are those?*
- *Who wrote that? Some old dead mathematician from England?*

The kids figured out the pseudonym of the person who wrote the riddle based upon their discussion of their own questions, along with a few more questions from me.

I then said that I had been online earlier today reading about the motivations behind the pseudonyms of Daniel Handler (a.k.a. Lemony Snickett) and Joanne Rowling (a.k.a. JK Rowling).

"How did you know that our conversation was going to lead to this topic?" asked E, shocked at my seeming prescience.

"She *told us* that she prepared the list of questions ahead of time," answered V. He was right. At the start of our Circle, I told the group that I had a few questions somehow related to the Pudding question that I was going to ask them.

"How did you know that our answers would lead to this point?" asked someone.

"Teachers *never* ask questions they don't know the answer to," replied V.

"I do sometimes," I replied.

"Like what," challenged the kids.

"Like one of the questions we've discussed today: *'Is it possible to lie without knowing it?'* I don't know the answer to that (Smullyan, Raymond)." Eyes bulged. Some students had never even fathomed that there might be something that their teacher doesn't know, or, moreover, that their teacher might admit it. How deliciously exciting and empowering. Exciting because they could see that I was not sitting in the All-Knowing-Teacher throne. Empowering because they realized that they weren't going to be spoon-fed what they were supposed to know and might be able to engage in meaningful work to figure it out.

My claim that the students realized they might be able to engage in meaningful work might seem a tall order, even a flight of fancy, but I think it's realistic. At this point in the class we had done a lot of back-and-forth in which the students asked questions and I didn't answer them. And despite stereotypes that exist in popular culture, kids want meaningful work that is fulfilling, means something, and is interesting.

I DON'T KNOW!

ROLE PLAYING THE BASICS OF LOGIC

The puppets came out now, and played a game of Knights and Liars.

"Waggy and Penelope are from an island where every person is either a knight, who always tells the truth, or a liar, who always lies," I explained. Raymond Smullyan devised this game, which he called "Knights and Knaves." It has become a piece of math folklore, commonly used to explore logic. In his book *What is the Name of this Book?*, Smullyan limited his characters to Knights, Knaves, and Normals. The additional characters you will read about, such as Blenders and Negators, are a creation of our math circle.

"But they're just acting," clarified the kids adamantly. "They're not REALLY knights and liars." This clarification is actually an important mathematical skill: Keeping track of which person was playing which puppet and which puppet was playing which role requires the kind of organizational thinking required to keep track of compound functions. (In these, the output of one function becomes the input for another. They can be taught via function machines.)

"Which one is a knight, and which one is a liar?" asked the students.

"That's what *I* was about to ask *you*," I said. "Listen to what they say and see if you can tell."

PENELOPE: "Jack is taller than John, and John is taller than Jack." Debate erupted over this (Smullyan) statement. Students attempted to resolve this seeming contradiction by addressing assumptions:

- "What if the second part of the statement refers to a different Jack and John?" challenged M. So Penelope refined the statement to specify that the identities of Jack and John remained constant throughout the statement.
- "What if the second part of the statement occurred after time passed and John grew taller?" challenged L. So Penelope refined the statement again.
- "What if they are the same height?" challenged D. This challenge was harder to answer. The kids discussed it independently, without my involvement at all. Many tried to convince D that this could not be true. They even demonstrated by asking kids of different heights to stand up. I faded into the background for some time.

Finally, with a couple of final questions from me, everyone was convinced that Penelope is a Liar because the "Jack and John" statement was not true. "Why is it not true?" I asked.

This was a tough question. The kids offered various restatements of their prior explanations. I was trying, unsuccessfully, to elicit a generalization. Either I was unable to phrase the question well, or the kids were not able to generalize at this age. Kids now started to doubt their conclusion.

"We have to make sure or else we'd be badly mistaken," said M, concerned. I suggested that we shelve this for now and move on to Waggy's statement.

WAGGY: "Charles Dodgson and Lewis Carroll are the same person."

After some quick clarifying questions, the kids agreed that Waggy is a Knight. The kids were projecting personalities onto them, and itching to take over. I was almost ready to abdicate. To deepen ownership first, though, I gave statements that reflected the children's personal interests at the time. I wrote them on the board and asked the children to determine who said each, Penelope (the Liar) or Waggy (the Knight).

- "Shane Victorino is on the Phillies" had one student waving his hand in the air.
- "Percy Jackson's father is Poseidon" elicited some jumping and shouting from several others.
- "Percy Jackson is a real person" led them to demand precision in the prior statement; they demanded I insert the phrase "in the novel."
- "An irresistible cannonball hit an immovable post" elicited heated student debate as though I weren't in the room. I considered leaving for a water break, but I wanted to hear their conversation about this classic Smullyan statement. When they finally came to a consensus that Penelope stated this, I gave 6 more statements.

By now, kids were still standing up, and again talking to each other, not to me. (Conversation that excludes the "leader" is what a math circle is all about.) At the end, I added even more drama by keeping score for the puppets—how many statements each owned. Our math circle is collaborative; the kids never compete with each other. But puppets competing? No problem.

After an hour of hard thinking, time was up for that week's session. I had used personal interests, pop culture, drama, puppets, competition, imprecise statements, and controversy to get into a delightful mathematical can of worms. Most importantly, the kids were providing more questions and answers than I was.

STRUGGLING WITH DEFINING MATHEMATICAL TERMS

Everyone learns differently. Says one reader, "Formal logic is a complicated topic even for an adult. When it comes to logic, I'm used to symbols without worrying about contradictions. As a person with a hearing loss, it's tricky — even if the words are heard — to comprehend what has been said and decide on the logic of it." Solving problems using words may be harder than solving with symbols for some people, while a symbolic approach is more intimidating for others. Working in multiple modalities insures that more students comprehend the discussion. Logic, for instance, can be taught through words, symbols, numbers, pictures, drama, geometry, proofs…In this particular course I used words and drama only.

The students' play revealed a gap in the rules of the game. I had to clarify a requirement on the Island of Knights and Liars: everything anyone ever says is a statement. "What's a **statement**?" asked someone.

"Something that's not a question or an answer," replied D. While the class quickly agreed that a statement is definitely something that's not a question, his reply led into an interesting debate about whether a statement could be an answer.

After the kids discussed and resolved that mini-debate, M demanded, "But what *is* a statement?" unsatisfied with the definition so far. Ideas petered out, so I gave one definition: a statement is a sentence that can be declared true or false. The kids then attempted to identify which of a number of sentences, including some commands, were statements. Things seemed clear until I announced that I was confused by what it means for something to be "**true**." M gave a definition that I refused to accept because it used the word itself in the definition. Now there was silence. I called on a few students for ideas, and got the response

"I don't know." So I did that thing that's so hard for us teachers: I let the silence hang there. We tend to want to rescue kids from intellectual struggle, which can be a disservice. Rescuing, though, can undermine the ownership that students get from true inquiry.

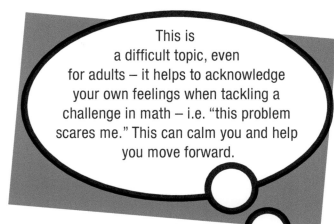

This is a difficult topic, even for adults – it helps to acknowledge your own feelings when tackling a challenge in math – i.e. "this problem scares me." This can calm you and help you move forward.

Finally, C offered the conjecture that true means "right." (Brave kid.) "And what does *right* mean?" I countered.

"Correct," said C. A number of heads nodded emphatically at his response. So I wrote "right/correct" on the board as our working definition of true. It was the best we were going to get for now. We all suspected there might be a better definition, but we didn't want the perfect to be the enemy of the good. Sometimes you have to accept what you have to move the mathematics forward. We can always revisit.

Why was I demanding such adherence to definitions? In math we use definitions not only to make sure we're all talking about the same thing, but also to meet a high standard of precision and mathematical rigor. Rigor in math means that our problem-solving follows a sequence of logical steps. Again, if someone can find a single logical gap in our sequence of steps, then the proof or solution falls apart. So we need to understand what logic is.

Term	Example	Shorthand	Informal Definition
Statement	It is raining.	R, U, or any letter	A sentence that can be judged true or false
Negation	It is not raining.	$\neg R$	A statement's opposite; negation turns true statements into false and false statements into true
Conditional statement	If it is raining then I bring my umbrella.	$R \rightarrow U$	An "if … then" relationship between statements, an implication

Term	Example	Shorthand	Informal Definition
Categorical statements	All umbrellas are shelters.	All U are S	A statement that gives information about membership in a set/category
	No umbrellas are cats.	No U are C	
	Some umbrellas are black.	Some U are B	
	Some umbrellas are not plastic.	Some U are not P	
Converse	Statement: If it is raining then I bring my umbrella. Converse: If I bring my umbrella then it is raining.	Statement: R → U Converse: U → R	A reversal of the two parts in a conditional or categorical statement; contrary to a popular misconception, whether the statement is true or false has no effect on the truth value of its converse
	Statement: All umbrellas are shelters. Converse: All shelters are umbrellas.	Statement: All U are S Converse: All S are U	
Inverse	Statement: If it is raining then I bring my umbrella. Inverse: If it is not raining then I don't bring my umbrella.	Statement: R → U Inverse: ¬R → ¬U	A negation of both parts of a conditional or categorical statement; only the first two types of categorical statements have inverses; whether the statement is true or false has no effect on the truth value of its inverse
	Statement: All umbrellas are shelters. Inverse: Some things that are not umbrellas are not shelters.	Statement: All U are S Inverse: Some ¬U are ¬S	

Term	Example	Shorthand	Informal Definition
Contraposi-tive	Statement: If it is raining then I bring my umbrella. Contrapositive: If I don't bring my umbrella then it's not raining.	Statement: R → U Contrapositive: ¬U → ¬R	A reversal of the two parts in a statement, plus a negation of both; if the original statement is true, its contrapositive is also true
	Statement: All umbrellas are shelters. Contrapositive: If something is not a shelter then it's not an umbrella.	Statement: All U are S Contrapositive: ¬S are ¬U	

THE EMERGENCE OF STUDENT-DRIVEN LEARNING

The kids were so emotionally invested in the role-playing at this point that they needed to operate the puppets themselves. Yes, needed.

I introduced the puppy, Wags, and asked the children to say any statement they could think of to him.

Now, I could have held Wags and said his lines/responses. The students didn't know what Wags was supposed to say. But the beauty of role-playing is that it gives kids power and helps them think.

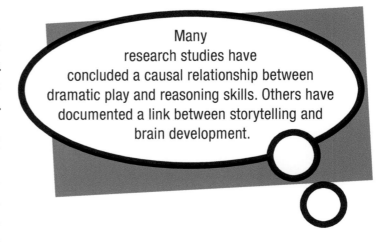

Many research studies have concluded a causal relationship between dramatic play and reasoning skills. Others have documented a link between storytelling and brain development.

I wanted them to own this, so I coached J. I whispered Wags' answers into J's ear, and J repeated them to the group. The students' statements for Wags flew out of their mouths so fast that my assistant (and co-author)

Rachel was not able to record them in her notebook quickly enough. But here are a few:

- "I am a cat," said V. ("You are not a cat," replied Wags.)
- "The blackboard is green," said D. ("The blackboard is not green," replied Wags.)
- "I don't like flowers," said L. ("You do like flowers," replied Wags.)

- "Jack is taller than John and John is taller than Jack," challenged D. ("Jack is not taller than John, and John is not taller than Jack," retorted Wags.)

Some of Wags' responses were true; some were not true. Student comments now shifted into conjectures and questions:

- "Wags is not from the island." (L)
- "Maybe he came from another island." (C)
- "Maybe he's half knight and half liar." (J)

- "Maybe his mother was a knight and his father was a liar." (V)
- "He was always saying the opposite of what we were saying." (D)
- "An Oppositer!" (M)

"What do you think Wags' favorite word is?" I asked.

"Not!" announced V, with E and the others chiming in. Several students challenged Wags to reply without using the word "not." The room was full of laughter.

"We say something and then he says the opposite of it," said V, thinking out loud. Someone else added that sometimes Wags lies and sometimes tells the truth.

"He's normal," said L, acknowledging our human tendency to sometimes tell the truth and sometimes lie.

"He's a **Normal**," declared V, formally dubbing Wags.

"A **Contradictor**," suggested someone else.

At this point I was giddy inside because I had neither introduced Wags as a Normal, nor defined Normals, nor ever used that term. Once again, I was reminded of the benefit of saying as little as possible. The students figured out for themselves the exact same concept that Smullyan had introduced in his book over 30 years ago! Moreover, they made this discovery (invention?) on their own terms, in a game that they had demanded and were leading. That is so empowering for the kids. I told them that yes, they were right, Wags is a Normal, a new type of person on the island. Normals, as defined by Smullyan and by our group, sometimes tell the truth and sometimes lie. Wags was a particular type of normal: a **negator**.

COMPOUND LOGICAL STATEMENTS – A BIG MATH DISCOVERY

"I love you and I don't love you," stated our newest puppet, Rooney the racoon.

"Liar!" It was obvious to everyone from the start that Rooney was a liar. I asked if he was a particular type of liar.

"He's a **blender**!" giggled M. "No one likes those people," she added, and the whole group laughed. I read the kids some sentences and asked whether Rooney, who always gives a statement coupled with its negation, could have said them. We used the students' term "Blender" for the rest of the course. Had I imposed some other name, like Contradiction-Stater or something else, it would have taken some power from the kids.

Since I stuck with the student-generated term "Blender," you may wondering why I did not call the "Negator" a "Contradictor" as the students did. I steered away from the term Contradictor because I thought it might confuse students with its ambiguity, as it could indicate someone who states either negations or full contradictions (a conjunction of two statements with opposite truth values).

This one, from Randy Mayes of Sacramento State University, stumped them:

"John Lasagna will be a little late for the party. He died yesterday."[1] This statement drove the kids into seeming chaos. "I like lasagna!" announced one. They were arguing, laughing, jumping, and shouting about it. Without a word from me, they paired off into lively independent conversations. V said to C, "Garfield ate John Lasagne, and Garfield was a little late to the party, so that meant that John was a little late since Garfield ate him!"

Things were digressing into silliness, but still on-topic. I couldn't get a word in edgewise. I did try. I had planned to discuss this example with the class, but as we were out of time and they clearly didn't need me, I decided to end class amidst the joyful noise. I had passed control to the kids. The kids easily finished off the John Lasagna question in the next session when V gave a convincing explanation of how a dead person could attend a party. Interestingly, not every student was convinced that we had finished that problem. "Is that THE answer?" wondered a few. They were surprised that some questions have multiple possible solutions. Moreover, I think they were surprised that they themselves had the power to debunk an apparent contradiction.

PUPPETS FROM HOME: PROGRESS VIA LETTING GO OF AGENDAS

"We want to use our own puppets here," the kids had requested. Bring 'em in, I said. Puppets from home led to kids inventing even more of their own math, which remarkably (or maybe expectedly) was some of the math I had on my own agenda. And despite having my own agenda, the kids were essentially running this course at this point.

What would you do if you were settled around the table with your students, about to engage in a civilized discussion about math history, when suddenly the students started chanting in loud rhythmic unison "Puppets! Puppets! Puppets!"?

It started because we had a new student that day. When several students and I started to fill him in on Dodgson's biography, a quiet murmur of "puppets!" could be heard. It quickly crescendoed. So we shelved Dodgson—why not? We moved to the other side of the room and got out the puppets for a game of Knights and Liars.

We reviewed our character categories—Knights, Liars, Normals, Negators, and Blenders—as I hung a piece of paper on the wall with that word written on it for for each category. We argued over whether Negators were a subcategory of Normals. The kids adamantly said no, so I bit my tongue, didn't correct or hint, and hung the Negators sign far from the Normals sign. Had the kids designated Negators as a type of Normal, the Negator sign would go below the Normals sign.

The game began.

> We moved to the other side of the room because it helps students to stay focused. Sometimes kids get distracted or just too excited, and this physical transition gives them something to channel that abundance of energy into.

- After E with Lammy the lamb, H with Leopard the turtle, and C with Waggy the fox had responded to statements the other students pitched them, the kids were directing the game. They told each other what to do and how to do it. I sat back comfortably with my legs stretched out, only assisting when asked.

- When S had Koko the gorilla respond to kids' statements, I quietly watched him and the group interact to determine that Koko was also a knight. I was thinking that this activity was not my agenda for today; I was ready to move into new material. I had planned to make sure we explore enough new material (Progress! Progress! Progress!) to ensure that we can solve the course's original mystery ("All Puddings are Nice") before our six weeks are up. But what is the overarching agenda of a math circle in general? The primary agenda of this one is to give a group of kids ownership of mathematics. So what the kids were doing here was actually pretty awesome in terms of that goal. I let go of my attachment to my agenda and watched the kids proceed.

> Why not correct or hint when the kids make a glaring math mistake? They will probably discover the mistake for themselves at some point, correct it themselves, and then own that math forever. If I correct it for them, they'll just get dependent upon me and relinquish some of their own problem-solving power.

- L chose Rooney the raccoon, and announced that he might not necessarily be playing the same role as he did last week. Students raised their hands and pitched statements. L made Rooney negate each one. Each negation was truthful, however, so Rooney's role was not clear. V raised his hand and said to Rooney, "You are a Negator." Now L hesitated. She didn't know what to make Rooney say. The kids thought about how to help her.

M gave her some advice: "Say you are not a Negator if you are one." (There seemed to be no need for an adult facilitator to give advice or ask clarifying questions at this point. The kids were doing it for each other.) Not everyone agreed with M's advice, though, so debate ensued.

Finally L asked me for help. She whispered in my ear that she wanted Rooney to be a Negator. I whispered back that she should make him say that he is not one, so she did. Through the ensuing debate, the kids realized that Negators are, in fact, a subcategory of Normals, so I hung the Negators sign below the Normals sign. The kids also realized (for themselves!) that they needed to ask more questions to ascertain Rooney's role. They still couldn't be sure which type of character he was. Play resumed.

"The sun is shining," said M to Rooney.

"No it is not," said Rooney. Everyone looked at the rain out the window and realized that Rooney was telling the truth. Therefore he could *still* be a Knight, a Normal, or a Negator. So people thought some more.

"The sun is not shining," said M to Rooney. L wasn't sure how to negate this one, so came to me for advice. I suggested that she have Rooney tell group that the sun *is* shining. Since my suggested statement didn't include the word *not*, she thought that she couldn't say it as a negation. I assured her that negations do not require the word not, but she preferred to have Rooney utter "The sun is NOT not shining."

Most students were now convinced that Rooney was a Negator. Two students, however, were not. Some people wanted to move forward anyway. I explained that we try to reach consensus when we can in a Math Circle. "May I give Rooney a statement?" I asked.

"No," said several students—only the kids could participate in this.

In this moment, our non-hierarchical structure was attained. Of course, I would have to continue to abdicate power to maintain this level of student ownership. So I asked whether the students would be willing to formulate a statement that Rooney would respond to in a way that would make his role obvious to everyone. No one wanted to do this—I think because they didn't want to have to think of such a statement at my command. Since the students felt secure in their power, however, they let their desire for assistance take priority over their need for power. They granted me permission to give Rooney a statement. "Rooney," I said, "you are a raccoon!"

"I am not a raccoon!" responded Rooney through L. Now everyone was convinced…

D brought his elephant, Buttons, from home. He asked the class for a statement, and had buttons respond with the exact same statement with the word order changed. "Are you a Scrambler?" asked V.

"That's really close," said Buttons through D. "But I am actually called a **Backwardser**."

Wow. My displaced mathematical agenda for the day had been to move into the **converse.** The kids, though, had arrived at the Backwardser—almost the same thing. It looks like we took a different route but ended up at the same destination anyway. We took a child-led detour and emerged in a stronger position. The puppets-from-home session deepened the kids' understanding of the concepts of true, false, negation, and most importantly, how to frame questions effectively. Their exploration has naturally led us to where we need to be. To get there, I had to allow the kids to take control.

I had
asked D to wait and go last
with his elephant puppet Buttons because at
the start of class, he had announced that his puppet was
a new category. I wanted all of the students to be secure in their
understanding of the categories we already knew about so that as many
people as possible would feel intellectually engaged in a potential challenge.
I was facilitating a raising of the bar from the floor upwards. Playing Knights
and Liars with puppets is an activity that might be called "low floor high
ceiling"; everyone can participate in a meaningful way without prior
expertise or experience, and everyone can advance it as far as
they want. Math circles are so much easier to experience
success with when the problems are low floor
high ceiling.

THE MATH GETS DEEPER AS TIME IS RUNNING OUT

The week after D introduced us to the Backwardser, I wrote a bunch of categorical and conditional statements on the board. My agenda: I needed to introduce these to get to the converse and the inverse to get to the contrapositive to solve the original Dodgson riddle. And time was running out.

In a really lame attempt to be clever, I brought in Waggy wearing a pair of high-top Converse-like sneakers. To introduce the converse, get it? Maybe not. The kids didn't either. But we still had fun and learned. I put the following statements on the board:

- *Baseball is a type of sport.*
- *Thunder is a type of storm.*
- *If it rains, then I get wet.*
- *If there's a fire, then there is smoke.*

Waggy gave the converse of each statement, for instance, "If there's smoke, then there's a fire."

The kids called him a Switcher, and a few other things. They described this manipulation as "taking the end and putting it at the beginning." I called him The Converser.

I called the character who responds to all statements with the converse of them the Converser and not, as one of the kids suggested, the Switcher, for two reasons: (1) there was no consensus on what to call him, and (2) there's an important mathematical distinction between the converse and the inverse and I was reluctant to muddy those waters with another term.

I asked the kids to predict Waggy's response to a few more statements, which they did with glee:

- *If I am a human, then I am mortal.*
- *If you are a sailor, then you are a good swimmer.*

I did not even have to ask for the truth value of each statement and its converse; the kids were calling them out. They were also coming up with their own statements for Waggy until everyone was shouting over each other—or, to quote Rachel, "chaos was on the brink of erupting." It was time to refocus, so I moved the group to the floor on the other side of the room and got out my bobble-head doll, which I used for a focusing activity. We ended the session with more puppets from home.

MY DECISIVE MOMENT IN THE FINAL SESSION

Then came our final (sixth) session. The kids were pretty disappointed. C was even fake crying. Unfortunately, 4 of our 9 participants were absent, which totally changed the dynamic of the group. M wanted to solve the "All Puddings are Nice" riddle, but everyone was on edge. D had done an exciting magic trick before class started, L wanted that trick explained, and V wanted to review and clarify what we had done last week. In other words, each student had a different agenda from the start. (Students have agendas too. It's a careful dance; as country singer Kenny Rogers once sang, "you have to know when to hold 'em and know when to fold 'em.")

The group discussed and re-enacted last week's roles, The Backwardser and the Converser. The converse was confusing; more was required than a simple algorithm. With Waggy the puppet playing the Converser role, the kids tried, unsuccessfully, to find a pattern in the effect of the converse on the truth value of the original statement. Then I pulled from my bag a dog puppet wearing a baby bonnet. The kids named him Baby Puppy. I asked the students to come up with some true or false categorical or conditional statements for Baby Puppy:

"All puppies wear bonnets."
"Puppets are pretend things."
"If it is a puppy, then it is young."

Baby Puppy stated the inverse of each statement, and the students easily grasped the concept. But then they began to argue over the truth value of the inverse. Several insisted, for instance, that the statement "If it is not a puppy, then it is not young," is obviously true. Others didn't agree with this at all. A few more examples led to the consensus, however, that the inverse is not necessarily true, but could be true in certain cases. (The distinction between must-be-true vs. could-be-true comes up again and again in mathematics, for instance in high-school mathematics such as algebra.)

The debate got out of hand when I had the puppets interact with each other. Baby Puppy the "Inverser" had responded to Waggy the Converser, who had first responded to a given statement. The result was the inverse of the converse: "If it is not young, then it is not a puppy." Students argued vehemently over whether the inverse of the converse of a true statement must be true. They also wondered whether the inverse of the converse is the same as the converse of the inverse.

We had almost achieved what was in my mind mathematical nirvana, the holy grail of third-grade logic, the **contrapositive**. (Boy, talk about an agenda!) Unfortunately, on this verge of mathematical insight, some students lost interest in the debate and engaged in paper-airplane making and chair tipping. The collective curiosity of the group vanished. So I moved the kids to the floor on the other side of the room.

Aaron Teasdale writes, "There is a decisive moment in any quality adventure when you choose to retreat to safety or leap further into the unknown." He was writing about a beach-cycling adventure in Alaska, but it applies to teaching too. This was my decisive moment.

My grandfather's take on this was that one should persevere, no matter what. (He was a champion athlete.) As a child, when I struggled with something very hard, he always said "When the going gets tough, the tough get going." But I knew that most young children do not have the inner resources to keep on truckin' when the questions get very difficult. We need to know when to take a break too. And when we're using role-playing, we have to be careful not to let the drama take over.

So here we sat, with half an hour left, and no future sessions. What to do?

WHAT I DID

On each of the boards I had written one line of our original Dodgson riddle:

LEFT BOARD: *All puddings are nice.*

MIDDLE BOARD: *This dish is a pudding.*

RIGHT BOARD: *No nice things are wholesome.*

I asked the students how Waggy and Baby Puppy would respond to these lines, and to each other. Unfortunately, since time was insufficient, kids were wild, and I was rushing. Or, maybe it's more accurate to say that I was rushing, and therefore the kids were wild. I wanted to get back to this problem since it was our initial mystery and there was student desire to solve it.

I realized at this point that six weeks was just not enough time for this group to solve this problem without leading questions. This topic is often taught in college, and these guys are only in the third grade. But I was also excitedly thinking, "You guys are only in the third grade! You are so awesome! C'mon dudes, you're on the verge of an amazing discovery! You can do it! Go for the glory."

Going for the glory: the question of whose glory it would be if these kids do solve this question is interesting to consider. Perhaps glory-seeking is another agenda I need to work on letting go of. I'm reluctant to generalize to all teachers, but I can say that I do have a bit of my own ego invested in kids' success in math circle. Therefore, we should consider what "success in math circle" means. My math circle is not achievement orientated, in terms of a product like test scores or math contest wins or big/numerous solutions to problems. But there is a process-oriented achievement goal: mathematical thinking in collaboration. For kids, both product- and process-oriented success is gratifying and confidence building. It is for us teachers, too.

I was really feeling for classroom teachers, who face this dilemma every day. Joe Strummer also came to mind:

"Should I stay or should I go now? Should I stay or should I go now? If I go there will be trouble, 'an if I stay it will be double."

I had come to class today solely prepared to use the puppets to tackle this problem. Since the puppet approach had been magical for the first five weeks, I thought of Bob Kaplan's advice to avoid over-preparing as I prepared for class today.

I came to class without my typical "Plan B" activities. So now I was faced with a choice: follow Bob's advice for when things descend toward chaos ("function machines!") or try to somehow refocus the kids on the problem.

I hurriedly put the puppet Waggy on one of my hands and Baby Puppy on my other, and told the kids that whoever can accurately predict Waggy's response to the first line of the poem could hold Waggy. (Yes, full disclosure, I took back some of the power by resorting to bribery.) Everyone got quiet and thought. Several students did come up with the converse to the first line.

I then asked for Baby Puppy's response to Waggy, in other words, the inverse of the converse, formally called the contrapositive. Everyone loved saying that word, but people got frustrated here because the second and third lines are hard to work with without rephrasing. We didn't have time to adequately discuss rephrasing. I was asking too much of the kids in too little time, so we did not solve the riddle.

Inquiry-based learning takes time. Rushing it, introducing extrinsic motivation, or emphasizing product over process can undermine it. I still don't know whether that was the right thing to do in this situation. I sent them home with the hint that writing each statement as if-then statements and using the contrapositive for help if necessary will lead to the solution. An anticlimactic end to this course, at best.

Spoiler!
You can use the contrapositive to solve this riddle. Another good approach is Venn Diagrams. Mark Paul and Sian Zelbo give this approach a nice treatment in their book Camp Logic. This visual approach would be especially helpful for people who favor visuals over words. However, the most direct verbal solution strategy may be to define each key part of Dodgson's riddle as a simple statement: P = it is a pudding; N = it is nice; D = it is this dish; W = it is wholesome. Then rephrase the original text as conditional statements, switching the order of the first two lines: If D then P. If P then N. If N then not W. This logical reasoning chain produces the final conclusion: If D then not W. In other words, If it is this dish, then it is not wholesome. Or this dish is not wholesome.

AND YET...

This was anti-climactic for sure. And yet, something about this course resonated with the students.

A year or two later, some of them were peeking in the classroom door and window as I let a math circle with younger students. I was using the puppets. At the end of my first session with the younger ones, several of the older ones came into the classroom individually and asked in betrayed tones, "Were *they* using the puppets? *What* were you doing with them?" And in curious tones, "What was the solution to that problem again?"

I wouldn't say. I gave them a hint toward the solution, and a reminder that a good math problem will revisit them repeatedly over time. It has revisited them multiple times, and I think that's due to more than the puppets and the quality of the problem.

When I let the kids take over the circle for six weeks, the mathematical logic stuck. My reigning in of the power at the end of the course is not what stuck. They carried with them the skills engendered by my mostly abdicating my power to student-directed learning. They directed it, ergo they owned it.

Human Rights in Math Class

Rachel age 14

This chapter deals with some very serious issues regarding human rights that may be disturbing, or even traumatizing, to some readers. This may be particularly difficult if reading about children who are hungry, tired, unsupported, or left behind in the school system may be upsetting to you, or cause you to remember or re-experience similar feelings from your own past. If this is the case, realize that this section may be difficult to read. There is no graphic language or "unsafe for work" content in this section.

Janet is a student in a large public high school. In her math class, she isn't always allowed to use the bathroom. Apparently she had asked to use the bathroom too many times, because one day her math teacher told her that it was becoming a "daily thing" and she wasn't allowed to go to the bathroom for a while.

Some of Janet's teachers don't let her eat in class. At Janet's school kids have lunch periods really early in the morning or late in the afternoon, so they get hungry during the day. Janet's lunch is at 8:30 A.M. One day in math class Janet had a terrible migraine from hunger but was afraid to ask her teacher to step out into the hall to get a bite to eat because the last time Janet asked, the teacher was really mad and barely let Janet go. If you eat in that teacher's class, you automatically get detention. So, Janet had a terrible migraine and couldn't concentrate throughout the class.

Now, Janet's story is heartbreaking. A poor student denied her basic rights in school. Every student has basic rights to: eat whenever they are hungry, drink whenever they are thirsty, use the bathroom whenever they need, and receive a good education. A child who is hungry and thirsty can't concentrate in class and has a compromised education. The same goes for a child who can't use the bathroom; their education as well as health suffers. A "good education" includes access to current textbooks and learning materials, a teacher who teaches and explains the material in a way that the student understands, and a teacher who assigns the right practice problems to the student to optimize their conceptual understanding of the subject.

When studying mathematics (or anything else complex), if a child isn't allowed to move their body, they can't concentrate. Moving the body is very essential to mathematics. Some people count on their fingers, and if they aren't allowed to do this, their ability to do mathematics suffers.

Rory is an elementary school student who uses her body to understand math. For instance, she counts on her fingers. When she started fifth grade, her new math teacher didn't approve of her counting on her fingers. Whenever her new teacher would notice Rory doing it, she would be quick to correct the girl. "Stop that! You need to learn how to count in your head, Rory."

Rory tried to stay still and count in her head, but she just couldn't. It got so that Rory started to sneak and count on her fingers under her desk. However, her teacher walked around the room to check on the kids and noticed Rory sneaking. Rory got in trouble.

Some people, like Rory, count on their fingers, and if they aren't allowed to do this, their mathematics suffers. When kids like Rory aren't allowed to move their bodies while doing math, they become restless and can't concentrate. Jo Boaler, Professor of Mathematics Education at Stanford Graduate School of Education, and Lang Chen, a postdoctoral research fellow in psychiatry at Stanford, argue for the use of fingers in mathematics. "Evidence from brain science suggests that far from being 'babyish,' the technique is essential for mathematical achievement … Schools across the country regularly ban finger use in classrooms or communicate to students that they are babyish. This is despite a compelling and rather surprising branch of neuroscience that shows the importance of an area of our brain that 'sees' fingers, well beyond the time and age that people use their fingers to count … Many teachers have been led to believe that finger use is useless and something to be abandoned as quickly as possible … Stopping students from using their fingers when they count could, according to the new brain research, be akin to halting their mathematical development. Fingers are probably one of our most useful visual aids, and the finger area of our brain is used well into adulthood."

In math class, when kids are denied their basic rights and are forbidden to move their bodies, what happens? When kids like Janet aren't allowed to use the bathroom in math class, they can't concentrate and do higher order thinking and tasks like problem solving. They are so preoccupied with needing to use the bathroom that they can only do mindless, rote tasks, such as counting or applying algorithms. They couldn't do something like create their own algorithms, for instance. They couldn't write a geometry proof. They couldn't do any problems that really required thinking. They could add numbers, of course, because they had already been taught an algorithm to do this easily.

When kids aren't allowed to eat they face similar problems. Just like Janet, they lose energy and become tired and drained, lacking motivation for complex problem solving. They have trouble listening and concentrating. They lose creativity because it takes extra energy to think in creative ways. When kids are hungry their bodies try to conserve energy; they have no extra energy to expend on creative thinking. They will not be able to do complex math problems or try to figure out new methods, because both of these tasks require creativity and energy. They will never develop the skill of mathematical thinking; they will only develop the skills of listening and following directions.

When kids aren't allowed to use the bathroom in math class, they can't concentrate and engage in higher order thinking and tasks like problem solving. They are so preoccupied with needing to use the bathroom that they can only do mindless, rote tasks, such as counting or applying algorithms.

When kids aren't allowed to eat they face similar problems. They lose energy and become tired and drained, lacking motivation for complex problem solving. They have trouble listening and concentrating. They lose creativity because it takes extra energy to think in creative ways. When kids are hungry their bodies try to conserve energy and have no extra to expend on creative thinking.

This denial of basic human rights in math classes is a very big problem. Curriculum designers are forced to design curriculums for students who are not able to concentrate, think complexly or creatively, and are unmotivated. When the curriculum does not teach deep thinking, that skill will never be developed in the students. It's a vicious cycle; because of the denial of basic rights in the classroom, students never have access to a quality education.

Some parents tell their children to sit still when doing homework to improve concentration. However, this does not work for all kids. Some kids need to think with their bodies, and when their movement is stifled, so is their analytical ability. Parents, if your children need to doodle, count on their fingers, sit in a certain position, stand up, lie down, etc., while doing math, this is fine! When kids are allowed to do math while being physically comfortable, they will be better able to think deeply and concentrate.

All of these stories we've discussed are about students without special challenges. It gets even worse when you add special needs into the equation.

Meet Brad. Brad, meet the readers. Brad is a middle school student who loves to move his body. He loves to play basketball and run track. He always wants to be active. He has ADHD.

At home, Brad has a small basketball hoop next to his desk and he likes to shoot baskets while he's doing math homework. His parents don't understand, though. They tell him to sit still while doing math to improve concentration. He tries to explain to them that shooting baskets helps him think, but they don't believe him. Some people, like Brad's parents, think that physical movement hinders learning and causes distractions.

However, modern brain research confirms what many of us suspected as we doodled in class: physical movement enhances mental processing. The idea that we can learn more efficiently through moving our bodies is called "Embodied Mathematics," or "Embodied Design." As science writer Colin Barras describes, "Researchers are discovering that learning is easier, quicker, and more long-lasting if lessons involve the body as well as the mind – whether it's gesturing with the arms or moving around a room … This theory is called embodied cognition, and it suggests that what goes on in our minds stems from our actions and interactions with the world around us. It means that encouraging children to think and learn in a purely abstract way might actually make lessons harder for them to understand and remember."

Some kids need to think with their bodies, and when their movement is stifled, so is their analytical ability. Parents, if your children need to doodle, count on their fingers, sit in a certain position, stand up, lie down, etc., while doing math, this is fine! When kids are allowed to do math while being physically comfortable, they will be able to think much more deeply and concentrate better.

Now, back to Janet's story. Janet's basic rights as a student and as a human were denied. Why did this happen?

The answer is complicated. First off, why would some teachers not let kids eat in class? In some situations, such as Janet's, the teachers don't want the kids to eat in class because the food attracts mice and rats. Previously there were janitors in Janet's school who cleaned up the classrooms at the end of the day. However, because of budget cuts in her school district, the janitorial staff was reduced. Now there aren't enough janitors to clean up after the kids if they eat in class.

Some teachers really go above and beyond and clean the classrooms themselves, allowing the kids to eat in class. For example, one of Janet's teachers lets the kids eat in class as long as they don't make a mess. If the room gets messy the teacher doesn't let the students eat in class anymore. However, it is extra work for the teacher to always be checking if the room is clean or not, and this shouldn't be expected of the teachers. Already the teachers are underpaid, have too many kids in the classrooms, and are perpetually worried about being laid off because of the budget cuts.

The more common reason teachers don't want kids to eat in class is because they think it might distract the kids. While it's possible that eating in class can be distracting, that distraction is minimal compared to the problem of trying to learn on an empty stomach. It's really hard for students to concentrate in class if they are hungry.

Chronic hunger among children is a serious problem; one fifth of children in the United States are struggle with hunger (Felling). Hunger can inhibit learning. "Kids with empty bellies find it hard to focus. They concentrate more on making it to lunch than on math or reading lessons. Test scores plummet (one reason schools feed kids a healthy breakfast on standardized test days). Hungry students can exhibit behavior problems; they may become irritable or rowdy or get lethargic. Some miss class to go to the school nurse, complaining of stomachaches and headaches" (Felling). Even less serious hunger can inhibit learning. "Even short-term hunger, common in children who are not fed before going to school, can have an adverse effect on learning. Children who are hungry have more difficulty concentrating and performing complex tasks, even if otherwise well nourished. Research and programme experience show that improving nutrition and health can lead to better performance, fewer repeated grades and reduced drop out" (Ponheary Ly Foundation).

The teachers are really doing a favor for both the kids and themselves by allowing students to eat in class in that nourished students perform better and understand more of the material. Of course if they are in violation of school policy and get disciplined or fired, that's not a favor to anyone. Moreover, I once saw a mouse in the room of one of the few teachers in the school who lets students eat. Eww.

In some smaller schools, all of the kids have a certain lunch period in the middle of the day. This way the kids have an opportunity to eat lunch in the middle of the day and don't need to eat in class. However, at Janet's school this wouldn't work for two reasons. First, at her school there are so many course offerings that all of the kids' schedules are different. It would be too hard logistically for all of the kids to have the same lunch period. Second, there are so many students in her school that they wouldn't all physically fit in the lunchroom at the same time.

Now, onto the next question: Why do some teachers not let their students use the bathroom? This is a harder question. It's possible that in the past teachers may have had students who asked to use the bathroom and took far too long, misused the time, or never came back. The teachers might have become distrusting of students in general because of these few untrustworthy students.

Alicia hates school, especially algebra. She always tries to sneak out of the class by asking to go to the bathroom. She has lots of things she loves to do in the bathroom. She loves to fix her lip gloss, text her friends, call her mom about a ride for later, catch up on homework for the next class, and talk with any of her friends who happen to be in the bathroom at the same time as her. She visits friends in the hallways, runs to the cafeteria to get some snacks, and sometimes runs to Dunkin Donuts to grab some coffee.

Her math teacher has stopped caring about her continued absences because almost all of the kids in Alicia's school skip classes. At least she doesn't skip class to smoke or do drugs like some kids might. Sometimes kids try to physically run out of the school. One middle school boy, Tom, was really angry at his teacher and yelled "I'm done. I'm out of here!" and ran down the stairs. The disciplinarian saw him running out of the school and ran after him as the kids watched out of the window. Tom ran as fast as he could and actually made it about half a mile into the surrounding neighborhood before the disciplinarian caught him.

However, I think the main reason why teachers don't want kids to use the bathroom during class is that they are afraid the kids will miss something important in class. They don't want their students' educations to be compromised. This is a valid fear.

It is possible that a student might miss something important by going to the bathroom during class. However, they could always ask their peers what they missed, or ask the teacher after class. I think that a student's education is compromised more by needing to use the bathroom and not being allowed to than by being allowed to use the bathroom and possibly missing something in class.

I think it is easier to fix the problem of teachers not letting students use the bathroom than the problem of teachers not letting kids eat in class.

Thinking about Janet's problems raises another question: Why do some kids ask to go to the bathroom and then abuse that time or fail to follow the rules? The simple answer is that the kids are bored or overwhelmed in class. Some classes are above students' levels, and when students get really overwhelmed and need a break to calm down, they go to the bathroom as a coping mechanism.

Some classes are below certain students' levels and the students are bored. This can be a problem in mathematics, where students can get stuck in certain tracks that are too easy. If a student has a less effective teacher their first year, they will be behind the next year; this has a domino effect that renders them less likely to advance into an "honors" or "advanced" section of math in following years. The end result is cyclic; the students end up in a math classes below their level every year. If the students get a less competent math teacher toward the beginning of their school career, they may never recover from that disadvantage and succeed in the subject, no matter what their potential is.

Now, why would some teachers not teach well? There are many possible answers. One could be that the teachers are just teaching their students the way they were taught in school. Pedagogical methods and neuroscience are always evolving, and staying up to date on current science and pedagogical methods can help teachers empower their students with a more valuable education. (Of course, some teachers can be amazing without looking at neural pathways—but staying current is never a bad idea.)

An important issue in the math classroom is pacing. Some teachers teach at a really slow pace. This is a problem if the students learn very quickly. Some teachers may teach at a slow pace because they think that by teaching slower they can go deeper, thus deepening their student's conceptual understandings of the subject. Teachers may also teach at a slow pace because they have large classes filled with children at different ability or experience levels, and have to teach to the children with the least experience. Many teachers may feel they have to teach at a slower pace to accommodate students with less experience or ability.

Sometimes teachers teach too fast and the students can't keep up. Teachers might teach too fast because they want to cover a lot of material in the course. "Breadth over depth is my motto. I plan to teach the whole textbook by the end of the year but not go very deep. I'll never do review," says one teacher. There are time constraints that prevent teachers from teaching at the pace that they think would be best for the class while also covering state mandated content.

We've talked about Janet, Rory and Brad. Now let's talk about me. One of the many problems with a coercive education is that grades become the main motivator. As my mom puts it, students "worship the Grade Gods." I have talked to many of my peers who feel as if most students at my school don't really care about learning, and mostly care about getting good grades. Kids also make sacrifices to the grade gods. One common sacrifice is sleep; many kids I know stay up all night to study. I try to balance the desire for grades with the desire to live, but it's hard.

In the spirit of Martin Luther's 95 Theses, I would like to formally announce my own theses.

- Kids should have the right to use the bathroom whenever they need to.
- Kids should be able to eat whenever they are hungry.
- Kids should have classes at the right pace.
- Kids should have the right to move their bodies in any way they want to assist their math learning.

Why Do People Hate Math?

Rachel age 15

Lots of people have had math ruined for them one way or another. In our culture, math has almost become synonymous with torture. I've heard people say that the word math means "Mental Awful Torture for Humans." If you mention math in conversation, someone will probably say "ugh," "I hate math," or "I'm so bad at math."

So, is math ruination inevitable? Will math always be that subject that hardly anyone likes or thinks they are good at? Or is there hope?

For many months I have surveyed people about how (or if) math was ruined for them. I've asked my friends and relatives, posted on Facebook, and asked teachers and parents. Even though everyone's story was different, there were some common themes that ran throughout. Therefore, I have grouped the responses into some broad categories.

Just to be clear, these are not *my* opinions of what ruins math; these are the results of an informal, anecdotal survey of many people of all ages, education levels, and attitudes. So, while you read these responses, I want you to think about something: can we fix these problems? Or are they unsolvable? Or can we transform the way that math is taught so that everyone can enjoy it?

PROBLEM: CLASSES ABOVE/ BELOW YOUR LEVEL

Having people of mixed math abilities in the same class, and having the class move too slow or too fast for you. I was once in a "regular" math class, and then after a few weeks moved into honors. In my regular math class we opened the textbook only once in almost a month. I'm not kidding you.

Then I switched into the honors class, and when I walked into the room the teacher announced "Chapter One test today!" and I was forced to take it. I got a 55. The two classes were at *such* different levels. It was almost impossible to switch.

It's really important that kids are placed in the correct math classes based on their abilities. It's a really big problem that kids are overwhelmed in classes that are above their level or bored to death in classes that are too easy. I know that in some high schools, once a student gets placed in a certain "track" (remedial, regular, honors, AP) it's really hard to get changed to a different one. So if a student finds that their math class isn't right for them, it can be almost impossible to change.

It's also a lot harder for kids to "move up a level" than to move down one. Sometimes the higher level math classes (honors/AP) move so much faster than the lower level classes that if a student stays in a lower level math class for, say, three weeks, and wants to move, the higher level class will already have covered so much information that it would be impossible for the kid to transfer into that class and get caught up.

Problem: Classes above/below your level

Advice (students): The conventional advice would be to talk to the teacher about extra help or more advanced problems. The problem with this advice is that I wouldn't follow it, and I don't think my friends would either. Adults always try to tell kids to talk to their teachers for extra help or more advanced problems, and, sure, maybe I've asked a teacher for help once or twice, but they're pretty unapproachable, not to mention busy.

And ask for harder problems? Never! School already takes up enough of my life - there's no way that I would want to spend more time on homework, even if you payed me. Easy A all the way!

Okay, time to be serious though. Once I was in a class that was way below my level, and I was bored to tears. So I got my mom to talk to the department head and convince him that I deserved to be in the honors geometry class instead of the regular one. So, in this case, I didn't ask for more homework; I asked to be switched to a class where the work was at an appropriate level for me. More of the wrong kind of work doesn't fix this issue, but getting the right kind of work does.

PROBLEM: TEACHER ATTITUDES

Teachers who make fun of and are unkind towards kids. In my research, I've found that a lot of people say that math was ruined for them because of "mean teachers." I was honestly surprised by this, because to be a teacher, you have to be a kind person who wants to make kids' lives better. What kind of system and working conditions could drive a benevolent and altruistic person to unkindness and a loss of temper? Obviously a terrible one.

Anyway, I'll tell these stories, with the proviso that *teachers in general are not mean*, and actually want the best for their students.

Once in a math class, when a student did not understand a concept immediately, the teacher pulled up on the projector a site advertising where to get jobs as a truck driver (implying that the student was only intelligent enough to be a truck driver). In another high school math class, when her high school students were whispering, the teacher screamed "Ladies and gentlemen, we are not in kindergarten anymore, learn to control yourselves!"

When my mom was in 2nd grade, she was absent for a day. When she came back, her math teacher said, "Go stand in the hall, Rodi!" (apparently a big punishment in those days). The teacher was mad at her because she was lost in class, but being lost in class was understandable since she missed a day of instruction.

Now, why would a teacher do things like this? I posit that the main reason is because of the teacher's unsatisfactory working conditions. Often teachers are underpaid and have to teach overcrowded classrooms. They live in constant fear of being fired if their students don't get high enough test scores. They are forced to teach a government-mandated curriculum, even if they feel that it doesn't do its job. Because of the strict curriculum enforcements, they are required to cover a certain number of subjects in a year, and are often cramming to fit the overload of information into the brief time slot (1-hour-a-day or less) they get.

It's really stressful being a teacher, and no wonder they sometimes lose their tempers. It's not really their fault. Anyone who had such a hard job would yell sometimes. I honestly can't imagine being a public school teacher, because going to public school I've seen these kind, creative human beings who just want to make kids' lives better become depressed, unhappy, and tired-to-the-bone.[1] All over the world, people become teachers because they care and want to make a difference, not because they want to torture kids. It's so sad that they aren't treated with more love and respect.

Although most teachers are kind, to become a teacher you have to be okay with having a lot of power over students. Teachers have the power to make students do things against their will because it will benefit the students in the long run. Teachers enter the profession knowing that there is an imbalance of power and have the responsibility of wielding this power peacefully. The issue of whether teachers/administrators should have the right to decide what is best for every child and subsequently force them to do it seems to be taboo. Most people, myself included, do think this is okay if it's for the kids' own good, and again, if the teacher wields this power peacefully.

Having teachers who aren't okay with saying "I don't know" or legitimately just don't know enough math. I think it's really important that if a teacher doesn't know an answer, they don't hide it and pretend to know. As a student, I totally understand that teachers are just human and don't know everything, and don't mind at all when they say "I don't know." In fact, I like it and I respect the teacher more for their honesty. It's important for kids to see that adults aren't perfect: the kids then feel better about themselves when they make mistakes of their own.

I have had teachers who pretended they knew the answer to everything, and it really upset me and the other students. We felt as if the teachers didn't treat us with respect when they ignored our questions that they didn't know the answers to. Kids and teachers can form better personal relationships when they see that they both are just human. When teachers and kids feel more comfortable around each other, deeper learning can occur.

So, why would a teacher be uncomfortable saying "I don't know?" Probably for the same reasons anyone might feel uncomfortable. Admitting you don't know everything about a subject you are supposed to have expertise in exposes you to the risk that your audience won't respect you. This may be true, although this hasn't been a problem in any of my classes. If teachers were more accustomed to just saying, "I don't know," and finding out the answers later when appropriate, a classroom culture of trust and respect between the students and the teacher might be more readily cultivated.

Sadly, in a lot of elementary schools, to become a math teacher you don't actually have to know that much math. Usually in elementary school one teacher teaches all of the core subjects. Some of these teachers actually have said that they don't really understand what math is. That's because they don't have to know that much about math to be certified. This is a problem. Math is introduced to kids in elementary school and that's when it's the most important that they learn what math really is.

Problem: Teacher attitudes.

Advice (teachers): Talk to your students about what math really is. I have had a few teachers talk about what math is at the beginning of the year, just as part of a lesson, and it has been very helpful for me. Most kids think that math is all about calculations and computations, and don't know that math is really about discovering the underlying structure of things, identifying patterns and breaking apparent patterns, proof and certainty, logic, and beauty. We as humans have this need for order and predictability and math satisfies that part of us. Math is about so much more than calculating and using algorithms. *What we learn in school is not beautiful, elegant math.* In my opinion the biggest thing that needs to change in schools is for teachers to start teaching real math, not cookbook math.

PROBLEM: PARENT ATTITUDES

Parents wanting you to be a genius at math. Or forcing math down your throat. A lot of kids face parental pressure to achieve in math. Most parents want their kids to learn, perform well on tests, get good grades, and in the end get into good colleges. This is all good and normal – of course parents want the best for their kids. They want their kids to do well in school so that they can have easier lives with minimal amounts of hunger, poverty, and suffering. Parents love their kids, and these desires for success are just ways to express that love.

The problems start when parents pressure their kids to achieve so much that it actually harms the kids. When kids are put under too much pressure to achieve, it can cause mental, emotional, and physical harm. Being under a lot of pressure can increase stress to an unhealthy level, and even cause physical health problems.[2]

As a kid, I know that a lot of my peers are terrified of bringing home their report cards to their parents in fear that their parents will think their grades aren't "good enough." Some kids I know face punishments if their grades aren't high enough—no phone, no computer, no going out with friends. Some kids also get rewards if they get good grades—new phones, laptops, money, trips, cars, etc. One kid I know told me, "My parents promised to give me $1,000 if I get straight As junior year."

It's easy to get sucked into this whirlwind of rewards and punishments. I get jealous of my friends who get all these presents, because I want stuff. However, the fact that parents give their kids rewards and punishments could show an unhealthy attachment to the kids' grades and achievements, which can make the kids really stressed out.

This question of rewards and punishments can be really hard. Here's a scenario: a kid doesn't work hard in school and only works to get good grades if their parents offer rewards for good grades. The parents say that they want their child to get good grades in school, thus enabling the child to go to a good college, thus enabling him/her to get a "good" job that pays enough to supply him/her with the basic necessities and hopefully a life of material comfort.

This scenario raises a lot of questions: Should the parents offer the child rewards/ punishments? What if the child doesn't care about a life of material comfort and wants to follow his own dreams? Is the child old enough/mature enough to make this decision for herself? What if someday there weren't grades anymore? Then what would be used to determine who "succeeded" and who didn't?

I'm not positing any answers. I think your answers to these questions really depend on your culture and values. These questions probably leave the realm of mathematics, but they are interesting to consider.

Posting grades/honor roll to Facebook. Social media can really increase the pressure on kids to succeed. I have friends whose parents take pictures of their report cards and post them on Facebook. This can be really embarrassing for the kids if they didn't get good grades. It's yet another mechanism to try to get kids to work hard and achieve. However, in my opinion, this tactic puts so much pressure on kids that they will probably achieve less than if they didn't have to be afraid of their parents publicizing their grades.

The same idea with honor roll – I have seen many social media statuses where parents said things like "Congratulations to my child, they made honor roll!" This can really increase stress and embarrassment for the kids. When kids get too stressed out, or live with a constant fear of not performing well enough, they can become emotionally and physically hurt.[1]

Problem: Parental Attitudes

Advice (parents): Try to pay attention to your child's happiness, health, and stress levels and be conscious about whether your actions are contributing in an unhealthy way.

PROBLEM: STUDENT ATTITUDES

Having students who do not ask questions in class. Very often in classes students do not speak up if they have questions. Sometimes, it's because they don't think it's worth it when they could just look up the answer in their textbook later. Kids are burnt out from so much schooling that they don't want to expend the energy to ask a question that they could answer for themselves later. This is true sometimes, but other times it's because of fear.

Kids are afraid. First off, kids might be afraid to appear dumb in front of their peers and teachers. If a student speaks up to say "I don't understand," they might be bullied or shamed because of it. Many times I have been approached by friends who said "I have a question, will you ask it for me?"

Although it is a harsh reality to accept, bullying is a very real and is a huge problem in schools. I'm not really going to go into detail about why bullying exists, but usually someone becomes a bully because they are not treated well and are not appreciated. Sadly, if someone is verbally or physically abused, they pass it on. And this passing on of abuse can often be in the form of embarrassing someone who doesn't know the answer. Aside from just being worried about what their peers will think of them if they don't understand something, kids are also worried about what their teacher will think of them.

Also, kids don't want to be that one person who slows everyone down. Teachers already have to teach classes with kids of mixed abilities, and it is hard for the teacher to slow down to explain things to the confused student while simultaneously trying to engage the advanced child. Students are afraid to be that one confused student who is holding everyone back, especially since they can sense that it might annoy some teachers--particularly if they seem to "always" be the person asking another question.

The reality is, if a student speaks up with a question, it's very likely that other students have the same question and just were afraid to say something. In my physics class, I would often cross my fingers that someone would ask the question I had; I didn't want to ask it and hold back the rest of the class. To really create an environment conducive to learning, the students have to be able to ask questions without the fear of appearing dumb and/or a burden. Currently, fear is dominating many classrooms, but hopefully someday it will be eradicated.

Problem: Student attitudes.

Advice (students): The conventional advice would be to talk to your teacher, parent, or counselor about your fear of speaking up in class. However, this is not what I would expect myself or any of my friends to actually do. Here's what I think the real advice should be: start out small.

If you've never asked a question in a class because you've just been too afraid of appearing dumb or holding people back, maybe just try asking one question a week. Then, once you're comfortable doing that, try increasing it to two times a week, and so on. Sometimes the best way to conquer your fears is to confront them in a controlled, safe environment. Little by little, the more questions you ask, (hopefully) the more comfortable you'll become.

PROBLEM: REWARDS AND PUNISHMENTS

Using bribery as an incentive for kids to do math (The Problem With Grades). In almost all schools, there are tests. Little tests, big tests, hard tests, easy tests. Aside from just having regular tests in class, teachers have to prepare students to take standardized tests. So much classroom time is spent preparing for tests that the tests actually hinder the learning.

Teachers have to spend a *huge* amount of time prepping kids for tests. Because of this tremendous focus on testing, it can sometimes seem like policymakers/administrators think that tests are more important than learning. Heck, even teachers get paid depending on how well their students score on tests! Schools get funding depending on how well their students score. Since tests matter so much, teachers have to find a way to make their students want to take them and score well on them.

Over time, grades have become a method of bribery to make kids want to take tests. If kids study and do well on a test, they get a reward – a good grade. If students do not perform well they don't get the reward. This system of grades can hinder learning.

If you had to choose between getting a "C" and really learning, or getting an "A" and not really learning, what would you choose? For me this is a hard question. I would like to believe that I would choose to get the "C" and actually learn, but I probably would choose to get the "A." Sometimes teachers criticize students for only caring about grades, but can you really blame them?

My little sister always tells me "All you ever talk about is grades!!" Students have been told their whole lives that if they get good grades they will get into a "good" college, get a "good" job, have a "good" life. They are told that if they do not study and perform well on tests they will have bad lives. This is a *very* strong message! *If you don't follow the rules and do as you are told, you will not have a good life.*

Being told this day after day, year after year, students learn to obey. Students become complacent with the fact that, sure, maybe they aren't learning a whole lot, but at least they will have good lives. (Remember that "good" really means "successful," and in our culture success is determined by achievement. This is an even larger problem. Students aren't taught how to have emotionally and spiritually fulfilling lives; they are taught to have lives filled with achievement and material possession.)

Students become afraid to rebel, to defy the system and try to learn. They are afraid to think for themselves. They become afraid of failure, an integral part of learning. They are afraid to ask questions. They are afraid to be creative and think outside of the box. They are afraid to live. This happens to me.

Problem: Rewards and punishment.

Advice (students): Try to remember that grades don't define you. Try to find time to lead a balanced and healthy life, and if this is impossible, think about changing your educational situation.

PROBLEM: NO ACCESS TO SUFFICIENT EXPLANATIONS

Not understanding what you are learning in math class and having no one to explain it to you. In math classes, the students often have mixed abilities. For some, the class is too easy, and for some, the class is too hard. For those advanced students, there is usually enrichment: for example the teacher giving the student extra problems, or maybe the student studying on their own. They can sometimes just do their own thing in math class and not bother anyone.

However, for the struggling student, there isn't an easy "fix." They can't just do their own thing in math class because they don't understand what's going on. Now, if you are a student who is struggling in class, what can you do?

Here are the usual answers: go ask the teacher for help, go to tutoring, or ask your parents. These are all great answers…except for the times when they aren't.

There are a lot of situations when the resources simply aren't available to help struggling students. Teachers are often overworked and do not have time to help a struggling student. In some schools tutoring isn't offered and the student's family can't afford to hire a tutor.

As for the suggestion of "going to your parents," there are many reasons why this isn't always possible. Some of my friends tell me, "I am *so* jealous of you! My parents don't speak English so they can't help me with my homework." A lot of parents don't know enough math to tutor their child, let alone understand it themselves. Many parents also work so often that they just can't devote the time to helping with school work, even if they really want to. I hope that someday all these resources will be available to struggling students, but the sad truth is that right now they are not.

Problem: No access to sufficient explanation.

Advice (students/parents): Try to find someone who can help you—maybe a parent, a teacher, a friend, a counselor, or anyone else. For example, if your math teacher can't help you, try approaching one of your math teachers from a previous year for help. If your parents can't help you, try talking to a friend who has a high grade in the class. If all else fails—or you just want to try this first—pay a visit to Professor YouTube. Say you don't understand completing the square. There are plenty of people who explain it well online. For students, I recommend Eddie Woo for videos and *mathisfun.com* for clear and concise written explanations. For teachers, I recommend James Tanton.

When kids don't see the point of learning math. Many kids I have talked to say, "I'm not going to use math in my adult life, so why do I have to learn it? It's not like I'm going to be a mathematician or scientist or something." One adult I talked to explains their frustration at not seeing a point to learning math: "I doodled and daydreamed my way right into remedial math. Which I also flunked. Then I went to summer school, where I double flunked. Some people are just wired for numbers. Others aren't. I wasn't. I loved words. Words make logical sense to me, whereas math was, and is, confusing gibberish."

"You can't force math to be interesting if your inner dialogue is screaming 'when am I ever going to need this in real life?!'" they continued. "And as a grown man with about a 6th grade grasp of basic arithmetic, I can tell you this: Unless you plan on doing something involving engineering, science or any kind of numbers reliant field, you won't be using most of the math lessons you are currently bored of. (I know. Not really a huge truth- bomb, but it had to be said.)"

They commented: "Not even basic fractions or algebra ever really come up. And hey, on the rare occasion that they do, calculators and the internet are totally things you are allowed to use for answers, like, all the time and nobody will ever think less of you for it!"

Many people share these ideas, and these people raise a valid point. In math classes where practical applications are introduced, they are usually *not* things that would actually come up in real life. For example, the ever popular "Two trains left the station…" Oy vey. Most people are *not* going to end up in a situation where they need to calculate train speeds!

Other than the basics, most people do not end up using the math that they learned in school when they are adults. Of course, the exception is if you go into a math-heavy field, such as engineering. So then, why do kids have to learn math?

If math is taught correctly, a certain way of thinking, "mathematical thinking," can be learned. Mathematical thinking is the way you have to think when solving math problems. This type of analytical thinking can be used all throughout your life—not just for math. It's not like you need to have memorized a certain formula to succeed in life; I mean, that's what the Internet is for!

The problem is that today in schools mathematical thinking is not always taught. If algorithms are given to students, if students only are required to memorize and not think, then mathematical thinking will never develop. The more you make mistakes, the more your brain grows (Boaler, Jo). If you are only given easy problems with algorithms for how to solve them, then your brain will never grow and your analytical thinking skills will not develop. Once schools prioritize mathematical thinking over speed and memorized algorithms, students will finally be able to struggle and really grow their brains in meaningful ways that will serve them their entire lives.

Problem: content and curriculum.

Advice (parents): If your child isn't being taught interesting math in school, you can help your child learn to love math by using resources outside of the classroom. There are lots of types of math enrichment: math circles, games, math clubs, YouTube videos, mathematical storybooks, and video games, to name a few.

PROBLEM: WRONG AMOUNT OF HOMEWORK

Having too little or too much homework. Your brain works like a muscle - the more you use it, the stronger it gets. So embrace the struggle! Working on hard math problems may seem frustrating and impossible, but in the long run your brain will get stronger and you will get better at math. That's why homework is really important; it's essential that you practice mathematical ways of thinking so that you can remember how to solve certain problems. The more you do math, the more analytical thinking skills you will develop.

On the other hand, too much homework is never fun. If you are coerced into doing math, and are forced to do it all the time, then any sane person would get a little bored of it after a while. Snooze snooze. I have to admit that I've fallen asleep my fair share of times while doing homework—it just gets so boring!

That's why it's important to find the right balance of having enough time to practice essential math concepts, yet also not spending so much time on math that you grow to hate it. Some studies have actually reported that there is an "optimal" amount of homework for each age; more than that amount of time is too much, and less is too little. It's hard for teachers to hit that sweet spot because of how much content they have to cover in a year; it's easy for them to assign too much, especially in advanced classes. However, assigning too little is also a problem and should be avoided.

Okay, here's a confession: I never memorized my times tables. I tried, I really did—I did some computer games where you had to navigate your way past trolls and things by correctly answering math questions ("What's 5 times 9?"). But honestly, I just didn't care enough. I always told my mom, "I don't *want* to practice my times tables!" I only ever needed about half of them in the math I was doing, so why bother learning the rest?

Throughout the years I've ended up learning most of them out of necessity, and it's been fine. Inquiry-based learning really does work. Sure, I didn't learn my times tables when I was younger, but I learned them when I needed to, so why does it matter? When a student really needs to learn an essential math concept, if they want to learn it and put in enough hours, then they can do it. Sure, some sweat and struggle will be involved, but in the end perseverance wins.

Problem: Wrong amount of homework.

Advice (teachers): Research the optimal amount of homework for each age and try to assign accordingly. I know that sometimes this can be impossible: you are under pressure to cover the whole curriculum, finish the textbook, and follow the Common Core Standards. Therefore, you have to assign a certain amount of homework to achieve your goals by the end of the year. If you would like to change the amount of homework that you need to assign, but are not allowed to, encourage your administrators to see the film "Race to Nowhere," which documents the ill effects of the wrong amount of homework.

PROBLEM: PHYSICAL CONCERNS

Being in an environment that is not conducive to learning. You know how some classrooms seem like the ugliest things you have ever seen and no one in their right mind should have to learn in there? If that's what you are thinking, then you are 100% right. Some environments are just no good for learning.

When I was little, I always remarked, "Mommy, that looks like a prison!" when I walked past my neighborhood public elementary school. If your math class is in a dingy classroom filled with mouse droppings and no light filtering in, then you might have the desire to fall asleep. This is especially true if all of the desks are facing one direction, so you can't even see most of the kids in your class.

Math is about discussion and cooperation! The best way to learn math is to work on it collaboratively. This is completely impossible if you don't even know the names of half the kids in your class.

One year in my math class, all of the desks faced in one direction. I hardly knew anyone in the class. Sometimes when my teacher would call on someone, the rest of the class would exclaim, "Who's that? There's a _____ in our class?!" If teachers want their students to be alert and enthusiastic, it helps to open up the windows, let some light into the room, move the desks around so that the kids can get to know each other, and offer some variety in the learning environment.

Of course I'm being idealistic; teachers are under *tremendous* stress and pressure to just make sure that all of the students pass and might not have the time or energy to make these changes. However, if possible it would be wonderful if sometimes classes could be held outdoors. Students with more views of nature outside of their windows actually score higher on exams (Matsuoka, Rodney). We as humans are connected to the earth, and when seeing nature we are calmed and our stress levels go down. All of my teachers throughout my years of school have closed the shades and turned on all of the lights. Maybe, if it's feasible, it would be nice to have some natural light sometimes. The benefits are enormous.

Last, let students move their bodies! It's not healthy for anyone to sit still the whole day. Moving our bodies provides us with energy and makes us more alert, which in turn will help us understand math better.

Problem: Physical concerns.

Advice (teachers): Try to move the desks around and let some fresh air and natural light into the classroom.

PROBLEM: FEAR

Fear is an overarching theme here. A parent told my mom that if there was such a thing as "Math Therapy," she would send her daughter. I was curious, so I googled "Math Therapy" and it's actually a thing.

There is so much fear surrounding math. Students are afraid that they will not know the answer, that they will be shamed in front of their peers. Students are afraid to be wrong and to make mistakes because this means getting bad grades. Parents are afraid that their kids will not do well in math and will fail, which could result in not getting into college and not having a successful life. Teachers are afraid that their students will not learn enough to truly understand math. Teachers are also afraid that they will lose their jobs, that their students will not score high enough on standardized tests, and that the resources will drain out and their students will not be given the attention that they deserve.

With all of this fear, it is no wonder that so many people hate math and are angry that they have to learn it. Underneath anger is often fear, and that is the case here: people are afraid that they will not be good at math and this fear is manifested as anger.

I want to help people become less afraid of math. Math can be beautiful and really fun to learn and do, if taught in the right way. I hope that we can get rid of all these problems, these "things that ruin math" and make math fun.

To do this, though, we need to get rid of the fear. The fear and the things that ruin math are connected. For example, a lot of this fear about "not being good enough" at math stems from the fact that we are scored on our math performance; we are judged. If there is less emphasis on testing, hopefully some of this fear about "not being good enough" will also go away. Math is wonderful for so many reasons and I hope that soon everyone, young and old, will be able to see this.

Problem: Fear.

Advice (teachers/administrators): Educate people about what math really is.

Once I had a math teacher who taught a truly amazing class. When I walked into the room, there were no desks and instead pillows, cushions, and beanbags all over the floor. There were lots of potted plants in the classroom, and outside the floor-to-ceiling windows we had a view of the woods behind the school.

We started the class (there were ten students) sitting down on the ground in a circle and, after doing a short mindfulness meditation, had a discussion about what we thought math was and if anyone had ever told us what it was. Then, on whiteboards we wrote down the different definitions and our questions. We talked about if any of us had math anxiety (we all did) and talked about why this was and how to reverse it. Then our teacher talked for a bit about what math really is and how it actually can be super fun.

For the first unit of class we went out into the woods and looked for the Fibonacci sequence in nature. Then, we collected examples and brought them back to the classroom to create an art exhibit. We also went to the art studios in the school and embellished our findings in any way we wanted while also creating new art pieces inspired by the natural world and Fibonacci numbers. Once we created all of our art we hung it all around school and also donated some to a nursing home.

The next unit it was time for food. We spent all week making delicious edibles so that on the last day of the week we could have a big feast. We learned all about fractions and multiplication by using single recipes to make food for the whole class (how much can one person *really* eat and how many times do we have to multiply the recipe to make enough food?) and also dealing with limiting reactants (we only have 4 3/7 cups of flour, but we needed 6. How can we adjust the rest of the ingredients to make our recipe balanced?). We had tons of fun making things like lasagna, pizza, brownies, vegetable soup, salad, pumpkin pie, garlic bread, lemonade, hot chocolate, samosas, pad thai, and many more things. By the end of the unit we had stocked all of the school fridges with delectables and had a delicious feast with only one kid throwing up from eating too much (Jimmy, and it was the cherry pie that did him in).

The next unit we spent in the school gym dancing. We had professional dancers come in to teach us to dance and also experts in the field of integrating math and dance. We learned all about concepts like symmetry, geometric shapes, patterns, multiplication and division, area, volume, and many more. We created our own dance routines and at the end of the unit we put on a huge dance show for the school.

Another unit we spent learning about math history through reading. Our classroom had a huge collection of books about famous mathematicians and famous unsolved math problems and we spend weeks just going to class every day, finding a comfy spot, and curling up with a book. After we spent a while reading we made a list of unsolved math problems that we wanted to try (we all secretly wanted to become famous for solving these super hard math problems). Then we spent awhile trying to solve them.

Anyway, you get the idea. We learned math in the real world. We went on numerous field trips to a variety of different places and saw how math is being used in the real world. We learned through exploring and following our interests in a learning environment that was all about collaboration. I learned more in that class than in all of my other classes combined.

Okay, spoiler alert: none of this ever happened. I totally made it all up! Could you tell? I've actually never been in a math class that's even *remotely* like this one. But, I wanted to make up what I thought would be the ideal math class. One that I would actually *want* to go to.

This class taught math in many ways, because people learn math in many ways. Something different works for everyone. This teacher recognized that everyone is intelligent in a different way and has their own insights to add to the class. Besides, who wouldn't go to this class—it sounds *so* fun! Who wouldn't want to play in the woods, create art, cook and eat food, learn how to dance, read a ton, and go on field trips all in one class?!

CAN WE FIX THESE PROBLEMS WITH MATH CLASS?

Remember, way back at the beginning of this chapter, when I asked you to think about whether we can actually fix the problems in math education? Well, it's time to talk. Obviously, there are a lot of problems in math education. It's quite easy to count on one hand the number of things that can be improved. I admit it, it's *way* easier to focus on the problems than the solutions. But just hear me out: the fact that we can identify the things that need to be fixed in math education means that *we can fix them.* If we know what's being done wrong, we can make it right.

And, believe it or not, there *are* ways in which math is being taught that make it fun and rewarding, and there *are* inspiring individuals working to reform math education every day. For example, look at the hypothetical math class I made up. That sounded pretty fun. And hey, if it didn't seem fun to you, don't sweat it; there are many ways to teach math, and they all work for a minority of people.

There isn't one "magic" or "right" way to do it that works for everyone. It would be awesome if teachers explored teaching math in tons of different ways, and some are already doing this, which is super awesome! If math is taught differently, kids can grow to love it and use it to discover deep truths about the world around them.

[1] In other countries, teachers are highly respected and have great working conditions, because in those countries education is a priority (unlike here) and it's very competitive and hard to become a teacher, because it's such a respected profession (the equivalent of doctors and lawyers here)

[2] If you want to learn more about how academic pressure can negatively affect health and mental/emotional well-being, I would suggest the documentary "Race to Nowhere" and Vicki Abeles' book "Beyond Measure."

How Not to Ruin Math

Rachel age 15

Okay, so we've already talked about how to ruin math—how to make it so awful that kids would rather eat Brussels sprouts than work on their math homework. There are so many problems with how math is being taught. It can seem depressing.

But, believe it or not, there *are* ways in which math is being taught that make it fun and rewarding, and there *are* inspiring individuals working to reform math education every day. There are many ways to teach math, and they all work for a minority of people. There isn't one "magic" or "right" way to do it that works for everyone. It would be awesome if teachers explored teaching math in tons of different ways, and some are already doing this, which is super awesome! If math is taught differently, kids can grow to love it and use it to discover deep truths about the world around them.

In these next sections, I've outlined some things that can be done to improve math education. Some are specific to math and some are more general, but necessary as well.

MATH-SPECIFIC CHANGES

Everyone is just trying their best to make sure students get the education they deserve—students, parents, teachers, administrators, and activists. Everyone has a different vision of *how* our students should be educated, but we all want the same thing: to raise compassionate, intelligent, and caring human beings.

Even if large-scale institutional changes can't be implemented, making changes in the classroom can really make a difference. It's really hard for teachers to teach math optimally when they have to follow a standardized curriculum, and it is almost impossible to ask them to try to make class interesting and engaging for kids when they have to worry about cramming in all of the information and preparing for standardized tests. However, I have met some teachers who do the impossible: follow the curriculum *and* make the lessons engaging and fun for the students.

However, these teachers sacrifice their health and their own time to do this. One such teacher always walks into school carrying an extra large coffee and is basically dead in the morning until he drinks it. He barely sleeps, has young kids of his own, is the sponsor of a school club, and is also active in the mentally gifted department at school (he's basically my favorite teacher ever).

Another teacher spends all of her free time grading; even at parties she is absorbed in her papers. I heard that any free second she has she is grading papers, but that's because she asked her first year Spanish students to write ten page essays. We also had to actually talk in Spanish—a shockingly rare occurrence for me in high school Spanish classes. She is my other favorite teacher, for obvious reasons.

It's really sad that to be so good they have to basically do their jobs 24/7. This is really depressing because it seems like with our current curriculum requirements teachers have to sacrifice everything if they want to engage their students. Hopefully the curriculum requirements will loosen up; if not, there are small things teachers can do without killing themselves to make math class more enjoyable for their students.

SHOW STUDENTS WHAT MATH REALLY IS

What do you think math is? Have any teachers ever shown you what math really is? I asked some people these questions. One girl I know from school said that "Math teachers never really tell me what math is, but science teachers always tell me that math is the language of science, and therefore math is the language of the natural world." The same thing has happened to me countless times! Science teachers *always* seem to tell the class that math is a tool for science, while math teachers don't explain what it is. Ironically, in my first high school math class the teacher was actually a science teacher (budget cuts, of course), and she told the class that math is important for communicating data for science. I actually had a conversation with her in which she said that her husband is a mathematician and he always gets annoyed that she doesn't see "math's beauty," but she doesn't believe that math has beauty.

Another friend responded to the initial comment by saying "Responding to what she said, I have to disagree. The natural world models free form, coincidental beauty. Math is based off of such a rigid structure, completely the opposite of the natural world." This is really interesting.

The first girl responded, "It's not that the natural world is based off of math, but rather the natural world can be broken down into math, which is why math is such an effective language for the sciences.

"The Fibonacci sequence can be observed in plants, nautical shells, and even the human anatomy. The very atoms and molecules that make up matter are bound together through electromagnetic forces that are calculable. Even the day and night cycles and seasons follow a form that's so predictable that many organisms have developed a natural cycle of their own, such as hibernation and mating seasons. Even our natural gravity can be broken down to numbers and formulas." Wow, interesting conversation, huh?

Another friend responded quite bluntly: "1) no, 2) a horrible thing," (meaning that no one has ever told her what math is and that she thinks it's a horrible thing). Along the same lines, another friend said that "math is the reason for the teardrops on my guitar." Some of us have been lucky enough to have teachers who really know what math is and are able to teach it, but others of us have not.

However, we can't blame teachers for purposefully tricking kids by teaching them cookbook math. That's crazy—although it might make a good science fiction novel: "Revenge of the Brainwashers in Math Class"! Teachers might 1), not have been taught what math is themselves, or 2), not have the power to teach mathematical thinking because of curriculum requirements. Sometimes teachers with really strict curriculum requirements have to "teach to the test" which can inhibit learning. These requirements can be really stressful for teachers and sometimes can cause them to assess students based solely on tests because tests are what "count" in the school system.

Actually, teachers can experience bullying, and as many as one in three teachers does.[1] Teachers are sometimes bullied by administration and coerced into following curriculums that don't teach real math. In an anonymous blog post, one teacher wrote that "I sat in the department meeting paralyzed by shock as the department head railed about an administrative 'crackdown' on nonconformist teachers. Glaring across the room at me, she said, 'We all need to be teaching the same thing at the same time. And people who don't like it need to *get out.*'"

In some schools there is so much pressure put on teachers to follow the curriculum that they can't teach what they want at all. Honestly, after going to public school as a student, I wouldn't ever want to be a public school teacher. I see the things that my teachers have to go through—the long hours, the low pay, the large class sizes, the pressure from administration, the lack of appreciation from students, the complaints from parents, and the rules against teaching what you want. It's just really depressing. But anyway, back to the responses to the questions.

One of my friends from school responded, "[My math teacher] has such a passion for math and learning. It's really inspiring; he defines it as everything, and I agree with that. It makes up every shape that we see—all the things we use in our daily life and how we can turn numbers and equations into shapes on grids and in real life—and I think math is beautiful. Math is the best." I totally love this definition and that she has had a teacher who has such a beautiful definition for math (I actually had the same teacher).

TECHNOLOGY AND GAMES: THE GREAT DEBATE

Whether adding games and certain types of technology to traditional math instruction enhances the quality of instruction is actually a *huge* debate. Should math education involve games, food (teaching math through following recipes), technology, manipulatives, and real-world applications? Or is math by itself enough? Meaning, does adding these "extra" things to math education to make kids actually care set actually reinforce the negative idea that math by itself is inherently boring and needs to be "made interesting" with games and other extraneous things?

Just think about it for a minute…okay, done? Have you decided? If not, don't feel bad. This question is really hard. I'm sure that it isn't black and white; not one of these ideologies is the one "right" one and the other one wrong. So let's just talk about it.

But first a disclaimer: I'm not talking about *computer math* here. I'm talking about *video game math*. Computers can be immensely helpful in doing things like simulating, performing huge calculations, modeling, data analysis, and more. They are a *huge* asset in the classroom. However, I'm not going to talk about them here because, as far as I know, there isn't a lot of debate about whether they are good or not. Most people I've met would agree that using computers to enhance math education is a wonderful idea. I want to talk about something that isn't as cut and dried. I wanted to dive into the debate and try to understand all sides. So, let's start exploring.

First off, many children already come into math classes thinking that math is boring and impossible and terrible and that they will never succeed. As one mathematician puts it, "Math is the most emotional subject taught in school. It puts us off guard. It makes us question our own intelligence." A lot of kids are simply terrified of math. This kind of math anxiety is often transmitted through and by parents. It can really be detrimental to the kids' math success.[2] It turns out that kids with math anxiety don't have as good of memory spans as kids who are more confident. They make more computational mistakes than the confident kids. These anxious kids come into math class expecting the worst. Maybe it is best to teach them math through games—to teach them that math can be fun. Then, once they see how awesome math is, if they ever have a desire to, they can transition into theoretical math, or they could do any combination of computational and theoretical math.

Math circle pioneer Bob Kaplan says that games and physical manipulatives should only be used in a certain way for certain types of kids. "Physical objects/manipulatives can be a release and help to get into the math. But the goal is to wean from the physical to the abstract. Hands off minds on." I completely agree (and I love the quote, "Hands off minds on!"). Kids who come into math class already hating and fearing it may benefit from games and manipulatives. However, once they have gotten rid of their fear and hatred, they can transition into pure mathematics, because as Kaplan describes it, games "may not be getting to the deep structure of mathematics."

So, if you're a teacher teaching a class of math-phobic students, or a student who hates math, what to do? Maybe try some math games! Or integrate math with some other love your have, like music, dance, art, or cooking. Personally, I did "math in the real world" until about seventh grade, and then I started doing some more theoretical math. I like both; I think it really depends on what mood I'm in. For example, in math circles I've made a lot of really beautiful art pieces that integrate art and math. I've also worked on purely theoretical problems involving chords in circles. They both have been awesome.

If you want to play some games, and also love technology, maybe you could try playing math video games. Some people are starting to think that certain video games can actually teach math! Well, not the type of video games that ask questions like "What is 3+4?" before you can move up a level (those can probably help you memorize your times tables though). Some types of video games are supposed to actually teach mathematical thinking. Some of the recommended ones are Portal, Wuzzit Trouble, and Minecraft.

Learning math through video games may seem like a natural way to learn math; in fact, research shows that a great way for young children to learn math is through play. According to Maria Droujkova, the founder of Natural Math, "Studies have shown that games or free play are efficient ways for children to learn, and they enjoy them. They also lead the way into the more structured and even more creative work of noticing, remixing and building mathematical patterns" (Vangelova).

Math video games seem especially beneficial for kids who already enjoy video games, and maybe not as useful for kids who don't already like video games. This was kind of my experience.

When I was younger, probably ten or eleven, I wanted to learn my times tables. It was really hard. So I tried some online games. I started to play "Timez Attack," an online game where I was a creature who had to navigate through a dungeon by answering times tables questions. I hated it. Then I eventually found an online "game" where it would literally ask this type of question: "6 x 12= ___?" I loved it. It was simple: it gave you a certain number of questions and then showed you how many you got correct. I loved trying to beat myself each time. (Actually, just a side note, I never ended up learning my times tables. I still don't know them. Once in Geometry a girl found out that I didn't know them and asked incredulously, "*What?! How can you survive without knowing them?*" Somehow I survived and am surviving to this day.) Anyway, I think the reason that I didn't like the traditional video game was because I've hardly ever played any video games. I'm sure that kids who love video games in the first place will find math video games fun.

Keith Devlin, a mathematician and video game inventor, tackles the question of the role math video games should play in the classroom. "If the technology had been available in 350 BCE, Euclid's Elements would have been a videogame" (Devlin as quoted by Sarkar).[3] Devlin thinks that video games are a good way to supplement math education in schools, but not as a way to replace teachers. He argues that a teacher does a far better job of teaching than a game does, yet games can help students see and interact with math in more than two dimensions and in a playful manner rather than just see and interact with it algorithmically with a pencil and paper. He thinks that this will revolutionize math education.

Video games have a two big things going for them. First, they are interactive. Second, they provide intrinsic motivation for correction. A video game tells you automatically if what you are doing is right or not, and if it's wrong, you are intrinsically motivated to correct yourself. When you learn something because you are intrinsically motivated, then you will retain the information for a longer period than if you learn something because you are externally motivated. It teaches students perseverance and doesn't punish them when they are wrong. It rewards work ethic instead of "intelligence." Studies actually show that if students are praised by being called "smart," they will actually be afraid to mess up, but if they are praised for being "hardworking," they will continue to work hard (Boaler). Video games will make the kids who work harder succeed and boost their self-esteem as well.

Devlin perfectly answers the question, "Why do we have to learn math?" His answer is that what we have to learn is mathematical thinking. Mathematical thinking is far more important than remembering some algorithm. This kind of thinking is taught through solving complex puzzles and problems, not having algorithms fed to us. These video games can teach real mathematical thinking (at least the good ones can).

Technology can enhance kids' math experiences[4]; it enhanced mine. My favorite things has always been to watch the videos that Vi Hart, a recreational mathematician, puts on YouTube. She makes these amazing doodles that illustrate really deep math concepts. My appreciation of math would be virtually nonexistent without watching her videos. Other technologies have impacted my math comprehension and enjoyment throughout the years, such as the use of smartphones in classrooms, different math modeling websites, and graphing calculators, but I have to admit that nothing even came close to how awesome Vi Hart's videos are. I really want to meet her someday! If you haven't heard of her, go check out her videos, they're really fun.

However, sometimes schools don't have enough money to purchase electronics for the students to use, and the students' families cannot afford to purchase them themselves.

THE POWER OF MOVEMENT AND COOPERATION

It's really hard for some kids to sit still at school all day. This makes sense, because studies actually show that it's unhealthy and unnatural to sit still all day (this totally freaked me out—I spend so much time sitting at the computer). Many kids learn best when they can move their bodies. It's actually really easy to integrate movement into math classes.

A wonderful way to teach math is *through* movement. Math and dance can easily be integrated. According to Malke Rosenfeld's website, "Math in your Feet," topics such as "congruence, symmetry, transformation, angles and degrees, attributes, categorical variables, manipulation and analysis of complex patterns, mapping on a coordinate grid, as well as deep experience with mathematical practices and problem solving" can be taught through dance. Dancing is a lot of fun and gets kids' bodies moving while they simultaneously learn math. No instructional time is lost. What better way to add more movement than by using it to learn traditional subjects?

It's also really important to integrate cooperation into math classes too, because math is best learned cooperatively! In math circles we work on problems together and have a lot more fun and discover far more than if we were solving them on our own. Everyone has different skills and knowledge. In the classroom, even if teachers aren't able to make their lessons cooperative, there are lots of easy ways to configure the classroom to make it more conducive to cooperation.

Moving the desks around can help students stay more alert in class and get to know each other better. Often all of the desks in a classroom are facing the same direction and the kids never really get to know each other. Math is far more fun for students when they can work together on problems; math should be about cooperation, not competition. In the real world, mathematicians cooperate instead of competing; school should be set up like the real world.

Another way to help is for students to be told to work together on math problems instead of on their own. In my Geometry class the teacher told us to work on our homework together in class, and it was a great system. If I didn't understand something, I could just ask the people I was working with to help me out, and vice versa. This eradicates the problem of kids being too afraid to ask their teachers questions; they can just ask their friends!

A circular arrangement for the desks is fun and advantageous, if the classroom is big enough and the class size is small enough. This configuration really encourages discussion. You can see the faces of everyone in the class, and this alone can make the atmosphere more cooperative and less scary. If possible this arrangement is an excellent choice because it rewards students with a much more interactive learning experience.

MATH AS AN ART - ADDING CREATIVITY

When I was little, I always thought that "math" was *so* boring. I hated basic operations, fractions, decimals, percents, and basic algebra. I would much rather have been reading a book or playing outside. I thought that math was rote, memorization-based, and non-creative. I liked to do things where I could use my imagination and creativity and think up new possibilities. Math seemed to lack everything that I loved.

When I started going to math circles, everything changed. In math circles we talked about the problems and worked together. We wrote lists of our questions, conjectures, and assumptions. It was all about trying to do things in new ways, using our creativity to come up with as many answers as possible, and having fun. At first I was dubious whether what we were doing was *really* math. "It can't be math, it's fun!" I told my mom. We watched videos and made art. We moved our bodies and danced. We read books about mathematicians and learned all about their personal lives. We asked deep questions:

"Was math invented or discovered?"
"Is math a tool? Or an art?"
"If you won't use math in your career, is there even a point to learning it?"
"Are there any *real* practical applications of math?"
"But does it *matter* if there are practical applications? If it's an art, why does it matter?"
"What is a line?"
"What is a point?"

And so on. We were encouraged to look at things from different perspectives and to think outside of the box. My thinking was revolutionized. All of my assumptions about math had been disproved. Something that I hated grew into something that I loved.

This issue of how *real* math requires creativity and an open mind ties back to the issue of school not teaching kids real math. Once schools teach real math, then naturally the lessons will be more creative, because math is all about thinking about things from different perspectives.

I want math to be like this for all kids. It's so magical when you discover something new that you love to do. I want to revolutionize the way that math is taught so that kids can learn and love math, that way that my mom did for me. If math can be taught through multiple media (videos, books, art projects, nature, movement, video games, puzzles and manipulatives, old fashioned pencil and paper, etc.), taught in an environment that encourages originality and creativity, and doesn't have an emphasis on testing and grades, math will truly become something that kids will love, learn deeply, and develop a passion for.

GENERAL CHANGES

There are lots of different things that teachers can do to improve math education in the classroom. There also are some systemic changes that have to be made so that every kid can have the best experience with math as possible.

NATURE AND THE RECESS ARGUMENT

It is *so* important for students to be able to learn in nature, or to learn with a view of nature. Sometimes I find myself saying "Oh my god, I could totally *live outside* someday! I just love nature so much that I would love to spend all of my time in it!" And then I realize: that's what humans have done for thousands of years. We are animals, and we are meant to spend most of our time outside. We aren't meant to spend most of our waking hours cooped up inside. That's why school is so hard for a lot of people. We want to be outside, we want to be moving our bodies, and we want to be interacting with the real world. That's how we really learn.

While it would be ideal to be educated outside, this is obviously a really drastic shift and will not happen anytime soon. It also is rather unrealistic for places with bad weather. A more manageable alternative is for students to have views of trees and nature out of their classroom windows. Just doing something simple like this can make a world of difference; studies have shown that having views of trees outside of the classroom windows lowers students' stress levels.[5]

Lowering the stress levels of students makes it far more fun for them to learn math. I can tell you from personal experience that it is basically impossible to learn when you are stressed out. I know that if I am in a class and have a test the next period in a different class, I will be too stressed out to pay attention in the class I'm in. Under normal conditions when I'm not so stressed out, I can actually concentrate on the present moment and learn.

What's interesting, and actually quite alarming, is that scientists have shown that stress hinders learning and actually damages the brain in areas that are crucial to memory, planning, self-control, and problem solving.[6] Many students face chronic stress because of an excessive focus on testing, grades, and homework. In fact, at least half of students at "elite" high schools experience chronic stress (Ossola). Many others experience chronic stress and even Post Traumatic

Stress Disorder (PTSD) due to issues that have nothing to do with school. Anything that can be done to alleviate this stress should.

Seeing nature from windows also has another positive effect: it can boost students' academic performance. The results of this study are mind blowing:

"Specifically, views with greater quantities of trees and shrubs from cafeteria as well as classroom windows are positively associated with standardized test scores, graduation rates, percentages of students planning to attend a four-year college, and fewer occurrences of criminal behavior."[7]

As you can see, even if you are not in nature, as long as you can see nature from where you are inside a building you will be less stressed out and can even do better in school! I know that sometimes it can be hard to change a view out of a window, especially in the inner city. However, just planting some trees or a community or school garden can make a world of difference to kids' mental health, happiness, and school success (Ozer). Another option would be to bring potted plants into the classroom or, if it's nice out, to have class outside. I have always dreamed about how nice that would be. I know that very often math teachers need to use whiteboards, but those are transportable and can easily be brought outside if desired.

Spending time in nature is also mathematical, because nature is full of math. It's so fun to look for fractals in nature; they really are all around us! Integrating the natural world into math instruction can help students see the beauty in math and not think that it is all theoretical. Also, looking at nature with a mathematical perspective can help students develop a greater appreciation for the world around us.

There's another thing that will reduce stress levels and help students succeed in math class: giving all kids access to recess! Recess has numerous benefits for kids, including increased focus, increased health, decreased stress, better social skills, and brain development, to name a few. All of these things are important for success in math class.[8]

All kids deserve to have time to play. Kids naturally want to play and explore the world around them interactively. Even when kids get older, such as in middle school and high school, recess is important. It has all the same benefits for older kids that it does for younger students.

In school kids mostly sit in silence and don't have time to socialize with their peers. Schools are replacing recess with even more instruction and standardized testing. This is going to get us nowhere. Kids need free time! Even teenagers are not mini adults—they are still children. They still want free time to move their bodies and socialize. All this can be accomplished by adding recess to the schedule.

Recess also has academic benefits. Kids learn so much from interacting with other children in play and games. Also, after moving their bodies and going outside the students will be better able to concentrate in class, will be more curious, and have more energy for learning. Even if math teachers are teaching math in amazing ways, this will not be very effective if the students are burned out from sitting still and being quiet all day.

In addition to all of these other arguments for recess, there is one more, which is perhaps the biggest: sitting down a lot shortens your lifespan. Sedentary lifestyles produce shorter lifespans, along with numerous other health problems. What's scary is that even if we exercise every day, if we still spend many hours sitting, this exercise may not increase our lifespans or reduce potential sitting-induced health problems (Schmid & Leitzmann).

Kids start this harmful practice of prolonged sitting at an early age because of school. Teachers routinely tell kids to "sit still," "stop wiggling," and "stop turning your head around." Teachers think that because kids can't sit still while they are in class, they are not learning. However, many people learn while moving, and aside from that, sitting in class is actually shortening kids lifespans. This really scared me when I read it. I took an online quiz to see how sedentary my lifestyle is and I was rated "highly sedentary" because I spend so much time working at the computer. Any change to the school day that lets kids move their bodies will be much needed and much appreciated.

REDUCE TESTING

Perhaps the most obvious way to make math education more fun is to reduce testing. When routine and standardized tests are added to the curriculum, kids learn less and become more stressed out and afraid to make mistakes, which in turn hinders learning.

Tim Walker, an author at the National Education Association, explains the devastating effects of standardized testing.

"Across the nation, the testing obsession has nudged aside visual arts, music, physical education, social studies, and science, not to mention world languages, financial literacy, and that old standby, penmanship. Our schools, once vigorous and dynamic centers for learning, have been reduced to mere test prep factories, where teachers and students act out a script written by someone who has never visited their classroom and where 'achievement' means nothing more than scoring well on a bubble test."

There are several problems created by our testing culture. First, too much emphasis is put on right answers, so kids become afraid to think outside of the box and explore new learning possibilities. This narrows kids' minds by forcing them to find one right answer to math problems, possibly stomping out the divergent thinking that is critical to mathematical success. Second, tests tend to be based on memorization, so students usually forget what they learned after the test. I have found that in math classes, when I memorize an algorithm for a test and don't conceptually understand it, I forget it after the test. Because teachers need to teach everything in the curriculum, they often cannot teach deeply and their students have to resort to memorization to pass the exams. This is not real learning. I have found that although I have taken many courses in school, I barely remember some of them.

Many teachers dislike standardized tests because of the way they hinder learning,[9] and this is already something that many people are fighting. More parents are opting for their kids to skip traditional standardized tests, because they take up so much time when actual learning could be done (Wallace). However, not all teachers agree on how big a role testing should have in their assessment of students, and whether or not there should be tests given in the classroom.

I was once sitting in the English office at my school making up a vocabulary quiz when I overheard an interesting conversation. One of the high school English teachers was in the office picking something up while another student was there as well, helping teachers grade tests in exchange for community service hours. (What's funny is that I was paying so much attention to their conversation that I almost failed my quiz).

"Wow, grading tests is really hard," said the student.

"I know, it's a struggle," replied the teacher.

"Why do you give them then?" the student questioned.

She paused to think for a few seconds.

"Maybe because we all are in one big rat race," she replied with a chuckle. "Or maybe because we don't know how to assess students without tests."

"Can't you use written assessments? Like describe the student's progress without tests?" the student argued.

"The parents and students expect us to give them tests, so that's what we do," she replied, as she grabbed her papers and left the office.

This exchange gave me a lot to think about. Why do teachers give tests? Is there anything wrong with giving tests? If they don't give tests, how should they assess students?

Alright, let's tackle these questions one at a time. Teachers give tests to evaluate how much their students learned. There's nothing intrinsically wrong with this, but there is a problem when preparing for tests becomes the only focus in the classroom. School is supposed to be about learning, not taking tests. When tons of tests are given in the classroom, going to school becomes all about preparing for the tests and not really learning.

This is true for many school subjects, but especially for math. On many math tests students are given problems which they are supposed to solve with an algorithm they have memorized. There is usually only one right answer. Students are discouraged from doing things like finding alternate ways to solve the problem and seeing if there is more than one answer. These things are really math. Math isn't only about using an algorithm to solve a problem, it's about thinking about the world in new ways and being creative and inquisitive. Classroom tests enforce an arbitrary practice that is disturbing because it's cookbook math with a focus on computation. Mathematical thinking, which involves *much* more than finding one right answer, is abandoned and the students are not learning. This is the problem with giving tests: they can hinder learning.

Now, a harder question: if we get rid of tests, how else could we assess students in math? Although it is not common, some teachers and schools don't have grades, like free schools. In those schools the need for assessment is eradicated. However, most schools do have grades of some sort, and want to keep them. So what should the teachers in these schools do to assess their students' progress?

Joanne Yatvin, a former teacher, principal, and president of the National Council of Teachers of English, writes that she didn't test students, and instead assessed them in different ways: "Our assessments of student learning were what any good teacher can see in students' projects, writing, artwork, talk, and behavior. They were not scheduled events, but observations of everyday activities."

As Yatvin describes, it is possible for teachers to assess more holistically rather than by a number of correct answers on a test. Teachers can look at stories kids wrote, plays, movies, books, and complex math problems they solved or worked on (Yatvin as cited in Strauss).

In mathematics specifically, homework, projects, and presentations of proofs could be graded. I'm not necessarily advocating that all tests be removed (ironically, tests arose in part to eliminate the bias that was seen in more subjective forms of assessment), but reducing our reliance on them as a way to assess students would be nice. Aside from testing, here are a few ideas on ways to assess students' mathematical achievement:

- Mastery of solutions to all problems assigned
- Demonstration of conceptual understanding of solution
- Coherent student explanation of own mistakes
- Successful identification of potential and teacher-dramatized mistakes

- Demonstration of ability to apply what student has learned to a real world context
- Demonstration of understanding of connections among graphical, numerical, analytical, and verbal representation of functions or problems
- Determination of the reasonableness of solutions

In math circles, we work collaboratively on math problems, often those that no one has ever solved. Students can learn math in so many ways: observing the world around them, playing video games, doing puzzles, shopping at the grocery store, and working on the kinds of math problems that mathematicians throughout all of history have struggled with. Teachers can observe effort; they notice when kids are curious about learning and work hard at things they are interested in. In a classroom that is more inquiry based, students will quickly develop their own interests and work hard because they care about what they are learning. Teachers can look at the projects they have done and the problems they have tried to solve. Teachers can listen to kids describing their work and can gauge understanding that way. They can pay attention when kids ask questions and have discussions about what they are learning. There are simply *so* many ways to gauge student understanding and work ethic other than testing.

Even if it is possible for teachers to switch to holistic assessments, the school would have to allow it. For example, during my freshman year of high school some of my teachers gave us projects instead of midterms and finals. In one class, we had a cooking project and my partner and I made Mongolian Butter. Some schools have a policy that for midterms and finals all teachers are required to give exams. If a teacher decided to assess students without tests, I don't know how the administration would react. It would probably depend on the school and the level of experience of the teacher. Also, teachers might not have enough time to do holistic assessments.

[1]"One teacher in three claims to have been bullied at work" (Southern Poverty Law Center).

[2]This study reveals how math anxiety reduces memory, causing other harmful effects: "Individuals with high math anxiety demonstrated smaller working memory spans, especially when assessed with a computation-based span task. This reduced working memory capacity led to a pronounced increase in reaction time and errors when mental addition was performed concurrently with a memory load task. The effects of the reduction also generalized to a working memory-intensive transformation task." (Ashcraft & Kirk).

[3]"When the big supertanker that is systemic education finally gets fully on board, and has seen enough examples of good educational video games to be able to separate the wheat from the chaff (currently it's mostly the latter that is out there), then I think that video games will play a significant role in mathematics education. Video games are a much better representation system for learning mathematics than are symbolic representations on a static page ... All Euclid's arguments are instructions to perform actions: draw an arc, drop a perpendicular, circumscribe the square, etc. It would be much more efficient, both as a communicative medium and for the student learning, if instead of writing instructions in words, the student was presented with opportunities to perform those ACTIONS" (Shapiro).

[4]The National Council of Teachers of Mathematics says that "It is essential that teachers and students have regular access to technologies that support and advance mathematical sense making, reasoning, problem solving, and communication. Effective teachers optimize the potential of technology to develop students' understanding, stimulate their interest, and increase their proficiency in mathematics. When teachers use technology strategically, they can provide greater access to mathematics for all students."

[5]"Even though the participants' self-reports do not show a relationship between the window view conditions and stress recovery, the two physiological measurements revealed a significant relationship between window views and stress recovery. The findings indicate that a natural window view has a stronger impact on stress recovery than barren or non window views" (Chen).

[6]"There's plenty of evidence that stress hurts cognition ... Research has linked chronic stress with the loss of neurons and neural connections in the hippocampus, a brain area that's crucial to memory. Loss also occurs in parts of the prefrontal cortex, which is essentially the brain's conductor, a key player in so-called executive functions such as planning, problem solving, and self-control," writes Vicki Abeles in her book "Beyond Measure" (Abeles).

[7]"In addition, large expanses of landscape lacking natural features are negatively related to these same test scores and college plans. These featureless landscapes included large areas of campus lawns, athletic fields, and parking lots. All analyses accounted for student socio-economic status and racial/ethnic makeup, building age, and size of school enrollment" (Matsuoka).

[8]"Everyone benefits from a break. Research dating back to the late 1800s indicates that people learn better and faster when their efforts are distributed, rather than concentrated. That is, work that includes breaks and down time proves more effective than working in long stretches. Because young children don't tend to process information as effectively as older children (due to the immaturity of their nervous systems and their lack of experience), they benefit the most from taking a break for unstructured play" (Pica).

[9]"Researchers have found that when rewards and sanctions are attached to performance on tests, students become less intrinsically motivated to learn and less likely to engage in critical thinking. In addition, they have found that high-stakes tests cause teachers to take greater control of the learning experiences of their students, denying students opportunities to direct their own learning" (Amrein & Berliner).

Function Machines

Rodi

Remember when the Nimbus 2000 came up in The Unicorn Problem chapter in this book? At that point, some of the 7-year-olds were rolling around on the floor and another student was flying on an imaginary "Nimbus 2000" broomstick as a few still persevered at a mathematical challenge. I needed to get the group focused, and quick. I quickly drew a broom on the board.

"Actually," I said, "the Nimbus 2000 is also a function machine. It makes this sound, "whoooooooosh," when you put something in, and then turns that something into something else."

Immediately everyone was sitting around the table, looking at the board, and totally engaged. "What part should I put a number into?" With my chalk I added a mouth right there. "Where should the result come out?" With my chalk I added an arrow over there.

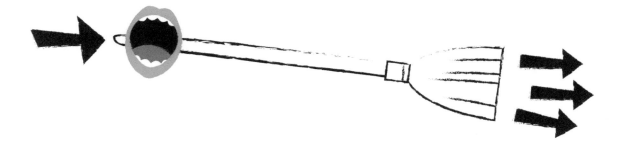

I "put in" 4 and "out came" 3.

"Put in 10!" demanded someone. Out came 9.

"Minus one!" the kids, age 5-7, announced, happy to have identified the rule. Function machines immediately suck you in because you want to know the rule. You need to know that rule. You just can't help it: you're human, so you have this desire to understand the underlying structure of the universe.

If you start small with this quest for understanding, with something like a rule for a function machine, you're that much closer to the world making sense. A world that makes sense is much less scary than one that seems arbitrary. Isn't that why our ancients came up with the gods? But I digress. Back to our story.

IN	OUT
4	3
10	9

"Give it some feet," requested someone.

"Louder!" demanded someone else. Now we had a new machine.

Our new machine, with feet and a louder sound, had a new rule. I solicited "in" numbers from the kids. In went 6, out came 3. In went 10, out came 5. This was much trickier. Finally someone posited that the machine "figures out what half of the number is and then minuses that from the number." Everyone agreed. In my mind I was dividing by 2, but none of these kids could multiply or divide. In the creative pursuit of math, there are multiple approaches to every problem.

A few children were shouting out the next few function rules before I was able to write multiple input and output numbers on the board, and more importantly, before most of the others had time to think. Once again, I needed to do something and do it fast. Typically when this happens,

1. I don't let them name the rule right away; instead, I ask those who know it to predict the next out number so that all students can catch up, or

2. I deliver a function that will really stump some of them. For simple functions, students only need to be able to count. Then you can raise the bar.

This chapter is the story of how I raised the bar with multiple groups of students over the course of a year. While we were at it, we playfully deepened mathematical thinking and moved toward abstraction with very little direct instruction. These students owned this.

RAMPING IT UP

"Add ears to the machine," demanded someone. Into this new machine went 10, out came 6. After a conjecture about "minus 4," in went 100, out came 51. Whoa. The kids were taken aback. After another conjecture was posited and disproved, in went 4 and out came 3, then in went 20 and out came 11.

"You keep changing the rule," accused P. Others concurred. I assured them that a function always does the same thing no matter what. The room was quiet; almost everyone had a furrowed brow but a hint of a smile. I gave a hint: the machine with ears does the same thing as the machine with feet but with one extra step. The hint did the trick; several students called out the function.

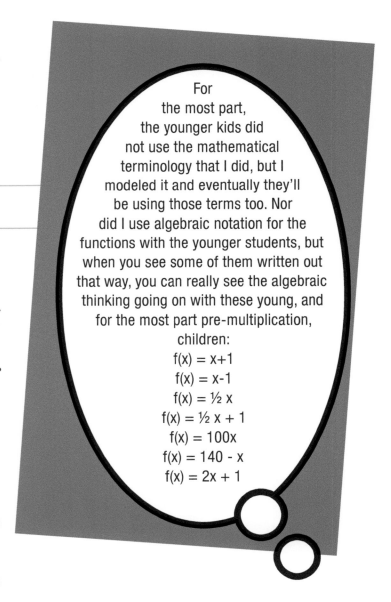

For the most part, the younger kids did not use the mathematical terminology that I did, but I modeled it and eventually they'll be using those terms too. Nor did I use algebraic notation for the functions with the younger students, but when you see some of them written out that way, you can really see the algebraic thinking going on with these young, and for the most part pre-multiplication, children:

$$f(x) = x+1$$
$$f(x) = x-1$$
$$f(x) = \tfrac{1}{2} x$$
$$f(x) = \tfrac{1}{2} x + 1$$
$$f(x) = 100x$$
$$f(x) = 140 - x$$
$$f(x) = 2x + 1$$

> I often hold up my hand in a "wait a sec" gesture to prevent the calling out of functions, but this one was just too exciting at this point – I couldn't get my hand up fast enough.

One student said, "I don't get it," so I demonstrated and encouraged that excellent strategy of counting on your fingers (and toes if needed). It appeared that everyone understood the function. "But you changed the rule," said M. "You can't change the rule."

Oops. I hadn't explained that every function machine in a round doesn't follow the same rule as the last one. I explained that when a technological advance comes along (such as feet or ears on a function machine, or a camera on a cell phone) a machine can perform a new task, or even a task involving multiple steps. The rule stays the same within each individual function; every input number will be subject to the same process as the prior one. But: revised machine, revised function.

> Ramping it up by adding in additional steps or more difficult concepts causes kids' brains to grow. Here I added an extra mathematical step, triggering a struggle; this struggle caused real learning to occur. The fun of function machines offers a nice foil for too much frustration. But can they be too much fun?

TOO MUCH FUN?

Children become exuberant in both their descriptions of the machines and in their desire to participate. It's important to know when to say when—and how. One day when a group's exuberance got out of hand, I lassoed it in with the lure of big numbers. People were at ease with basic adding and subtracting, so I put up a function that produced 200 from an input of 2. Several kids discharged their energy by competing with bigger and bigger input numbers. "Two million three thousand and eight hundred!" declared someone. "2,003,800," I wrote on the board in the "in" column of our table. (Most students were surprised by what this number looks like.) Then for the "out," 200,380,000. The kids wielded bigger and bigger numbers:

> Behavior challenges are a normal thing.

"5,500,170!" (Take that!)
"99,099,099,999!" (Ha!)

IN	OUT
2	200
2,003,800	200,380,000
5,500,170	550,017,000
99,099,099,999	9,909,909,999,900

No one had any clue what the rule might be. I was adding two zeros to each number, but the input and output numbers looked scary and incomprehensible.

"I have a good idea," said A enthusiastically, "why don't we break up into teams and try to figure it out in small groups?" I agreed that this was a good idea, but unfortunately an idea we didn't have time for today.

"Do you think we can solve the function if we keep putting in such huge, unfamiliar numbers?" I asked.

I didn't mention this, but with the high energy level in the group, I thought small groups would not be able to stay focused. Who knows if I was right? Also, I tend not to group students in order to maintain the tone of collaborative conversation. If I had a larger class and a few additional adults to act as administrative assistants, grouping would work well.

"Let's try 1," suggested D, seeing some logic in this implied suggestion. When 1 produced 100, people were still mystified. We tried some other small numbers.

"So, when you put in 1, you got 100. With 2 you got 200. With 3, you got 300. What if you put in 4?" I asked.

"You get 400," answered everyone. Everyone was accurately predicting outcomes, but no one could name what was happening.

"What if you put in 11?"

This was met by silence. I started to suggest another in number when P said, "Wait a minute. I think I know. You would get one thousand one hundred." He was right on the money, but no one else understood. I asked if anyone knew another name for that number. No. I asked them to help me count by hundreds: one hundred, two hundred, etc., etc., nine hundred, ten hundred… "Oh!" shouted someone, "It's eleven hundred! The function machine makes hundreds!" Everyone agreed.

I asked if anyone knew how to make hundreds mathematically, and no one did. I had ramped things up by introducing a new concept (multiplication). Everyone was very happy to declare this function machine a "Hundreds Maker," which it certainly was.

> There are multiple models of multiplication, including both scaling and stretching. This is deep conceptual stuff, which we arrived at with some fun. This is a nice jumping-off point to begin to explore multiplication. I didn't do that here because the course was on a topic totally unrelated to function machines or number theory. I was primarily using them as a focusing tool for and mathematical relief from a totally different topic.

MOVING TOWARD ABSTRACTION

Another day, another dollar. Wait, no: another day, another (slightly older) group, another machine. As I drew the machine, these 8-10 year olds dictated its parts: various polygons, a cupcake, and sprinkles. They specified that the "in" number should enter the right eye and the "out" number exit through the cupcake. I put in the first number: 70.

The students corrected me when I put the "out" number in the wrong eye. "When you put in 70, 70 comes out." Faces twisted in thought. Then, 90 in 50 out. 40 in 100 out. The total silence of very hard thinking. Then the kids started thinking aloud and producing conjectures and follow-up questions:

- "As the in numbers go up, the out numbers go down."
- "Maybe they are negative numbers."
- "What happens when you put in 1?" (139 comes out.)
- "What about 100?" (40.)

More silence.

I threw them a bone: if you put in 2, 138 comes out. A few faces lit up. M noticed that "each pair adds up to 140!" All agreed enthusiastically, confident that they had solved the puzzle of what the machine was doing.

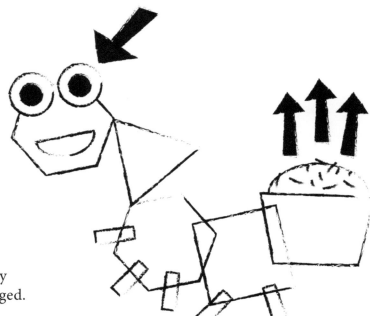

But I didn't let them off the hook so easily. "What is the machine doing to the numbers?"

"I don't know," said a few aloud. Every mind was one-hundred percent engaged.

Then D said tentatively, "If you put in a number, it finds another number where if you add it to the number that you put in, you get 140." I asked him to repeat his conjecture. He did so, with more confidence, as other heads nodded in agreement.

"I think you're right, but it's hard to keep track of all those words," I asked. Is there some math operation word that could describe this in fewer words?"

A few kids suggested addition, but were unable to say what exactly was being added. Finally M asked, "Could it be subtraction?"

"What would you subtract?"

Several voices excitedly joined M in declaring, "Subtract the number from 140!" The rest of the group audibly exhaled in relief.

I wrote the rule on the board: "subtract from 140." This rule requires a lot more thought than, say, "subtract 140." This was a very hard problem, but collaboration and perseverance made the solution possible.

Interestingly, the students had a much easier time with this two-step addition machine than with the one-step subtraction machine. Subtraction seems a little more abstract than addition. It is taught in arithmetic as a separate operation from addition. Yet in algebra, which is considered substantially more abstract than arithmetic, subtraction is seen as the addition of the opposite. What's the opposite of, say, 5? Negative 5. The person in our circle with the conjecture of negative numbers was on the right track, but didn't know how to verbalize it. I wonder what would have happened had anyone suggested an in number larger than 140 – that would have generated a negative out number.

I once was tutoring a high-achieving high-school student who had already completed a year of calculus. We were reviewing algebra II for a college-admissions test. "What is a function?" I asked her. "Something that passes the vertical line test," she responded. "But why does it pass that test?" She didn't know. She may have learned math via a cookbook approach, or she may have simply forgotten some basic algebra since it had been a few years. Either way, sometimes things can become so automatic for us that we forget what's really important.

The next machine had an elephant where the cupcake had been, and used the function 2x+1. Of course, these young students did not express the function/rule algebraically once they figured it out; their wording was either "add it to itself," or "double it."

INTRODUCING MATHEMATICAL CREATIVITY

All of my students want to challenge each other with self-created function machines. Actually, I do too. We want to create something that builds excitement and suspense as people try to discern the rule. Suspense verifies for us that the rule is important. That our thoughts and creations are important. If people put 100% of their attention on our function machine, we must have created or discovered something of value.

While we're at it, let's define a function here: a function is a rule resulting in only one output number for each input number. In other words, only one y for each x; hence, if you draw a vertical line when you graph the function, and the vertical line crosses your function more than once, it's not a function. Or, for each in number, you can only get one out number. $y=x^2$ is a function, but $x=y^2$ is not. It's just a relation.

We don't, however, want to frustrate people to the point of giving up. We put ourselves into what we create. We want to be seen, to be understood, to be understandable. We want to connect with others. So a creator of a function machine walks a fine line: don't make it too easy, but don't make it too hard.

One day I told my students that next class (finally!) was their turn to present their own function machines to the group. They had been begging to do this for weeks, but we didn't have the time. Now that day was upon us.

D, the youngest, asked, "Am I the only one who brought a function machine?" He held up a large tote bag and announced that "modifications" could be added as needed. When others started suggesting input numbers, he realized that he hadn't fully developed a rule. He decided to sit down, think about it more, and produce his later. Then some other students presented theirs. Turns out everyone had brought in a function machine, just without props.

The functions that people had prepared at home generally involved either digits or parity. A collaborative approach was definitely needed to solve these machines:

- N's machine: if it's odd, double it, and if it's even, triple it
- G's: the units digit of the out number is 2 more than the tens digit of the in number, and the tens digit of the out number is 3 more than the units digit of the in number
- M's: a more complicated variation of G's machine

At this point, no one was calling them "machines" anymore. Almost immediately, the kids suggested that we not spend time drawing the machines, and just get to the rule. So what we had been calling "function machines" when I presented them were now simply "functions." We spent some time discussing the difference between 1-step/2-rule (conditional) functions such as N's and 2-step/1-rule functions such as G's. Then K challenged the group with a function that first adds 8 and then takes away 3. The kids had fun debating whether functions could be equivalent and seeing explicitly how functions can be simplified.

And then D was ready with his functions:

- if you put a number in the bag, the number that came out was 3 more.

Then he modified the bag by inserting a toothbrush to create a new rule, which appeared to the other students to be:

- "if it's odd, add 1; if it's even, add 2."

D did not accept this as the rule, though, because he remembered from a prior session that conditional functions could sometimes be simplified into 1-step rules. He did not accept the verbalizing his rule as a 2-stepper. He demanded it in one step:

- the next larger even integer.

Wow! Every time I lead a math circle, it seems that I unwittingly form expectations, and that they are usually debunked. Here, our youngest student had defied my expectations about age and math by presenting the most challenging function.

WHY ARE FUNCTION MACHINES SO RIDICULOUSLY ABSORBING?

1. This is creative work. When D put his toothbrush into his tote bag and designed his new rule, he was discovering within himself what Peter Korn calls "the capacity to transform a wisp of thought into an enduring, beautiful object." It's not uncommon for people to describe their work in mathematics as quests for simplicity, eloquence, and meaning. But it's not just the product that matters; I posit that the process is even more important. Our minds and wills are totally engaged in what Korn calls "a place of creative challenge and grace that has nurtured a reliable sense of meaning and fulfillment throughout my adult life."

2. We crave meaning and structure. We create to understand how the world works. Of course it's fun, but something deeper is going on. "However minutely we may be absorbed in our own 'stuff'," says Korn, "through creative practice we investigate existential questions such as 'Who can I become?' and 'How should I live?'" We want to see things for what they really are – in math and in life. In life we want our own inner light to reveal itself after we peel away the layers like an onion. We assume that there is an inner light. In math, too, we assume that there is an inner light in the form of an underlying structure.

3. We might use mathematics not only to understand ourselves, but also to transform ourselves. When students N, G, and M set to work at home on their functions, they set out to "bring something new and meaningful into the world." They manipulated their medium (mathematics) "in previous untried ways to tease that meaning into being. The result is a novel, first-person experience that, inevitably, redraws" their stories about "who one is and how the world works. . .Creative engagement is an experiment through which the maker seeks new ways to envision human potential, using himself as the laboratory." (Korn)

This is heady stuff. Is it true? I don't see why not. By the way, Korn is a professional woodworker. His medium is wood and his tool a lathe. My field is math education. Does that matter? I think not. I suspect there are many endeavors that can serve us in our desires to be creative, seek meaning and structure, and to transform ourselves. Or, to quote Rachel, "math is inherent and our drive for meaning in our world is something that is expressed through our desire to solve math problems."

ALGEBRA VIA FUNCTION MACHINES

In goes 2, out comes 4. In goes 3, out comes 9. The rule was obvious to a group of 10-12 year olds, but the terminology was not. "I never knew that little number was called an exponent," said R. Other terms that a handful of kids knew were "squaring a number," and "taking it to the power of 2." *For now*, we did not use variables to describe the functions, and instead used operational symbols with the words "in" and "out". So we described "$y=x^2$" as "$out=in^2$"

I raised the bar: in went 1, out came 2; in went 2, out came 4, in went 3, out came 8 ("what?!") and so on. "Is it Fibonacci?" asked M. (No.)

"I think I see a pattern in the out numbers: they keep doubling," said P. Faces lit up, but I burst that bubble by switching the order of the pairs of "in" and "out" numbers to demonstrate that this function is not recursive.

"This function totally depends upon the number that goes in," I explained.

After a lot of staring and thinking, and the hint that this function was similar to the previous one, L noticed that the in number dictated how many 2s there were in the out number. Then M sat straight up and announced, "I get it! You make the 'in' number the exponent!" With a little more work, the students (not me!) worked out that the function was "$out=2^{in}$" ($y=2^x$).

INVENTING ALGEBRA WITH
MORE FORMAL NOTATION

The students deduced three relationships between input and output numbers in the next week's function machines. I wrote out each relationship on the board in full sentences. "My hand is too tired to write out the rules in words any longer," I whined. In response, they collaboratively constructed equations using the words "in" and "out" for the variables:

"You multiply the *in* number by itself."
"You multiply 2 by itself the number of times the *in* number says to."
"You multiply 1/2 by itself the number of times the *in* number says to."

$$out=(1/2)^{in}$$

$out=in^2$
$out=2^{in}$
$out=(1/2)^{in}$

These came easily once we reviewed the definition of an exponent. For the next function, the students quickly supplied the equation *out=in times 2*.

"But can we get rid of that word *times*?" I pleaded.

Kids first suggested *in x 2*, and then remembered that we can assume multiplication when writing equations, and simply write *in2*. "There's a convention in math, though, about the order," I began. Immediately, L raised his hand and suggested *2in*, instead of *in2*. "So, in math, the convention is to write the thing that does *not* vary to the left of the thing that *does* vary. By the way, in math, there's a name for the thing that varies."

"A variable!" chimed in the few kids who are already doing some algebra.

"So," queried M, "why don't you just write *i* instead of *in*?" I did.

"But that's the square root of negative one," replied L. Most faces looked seriously perplexed, as they had never heard of such thing. We let it go, though, because it was not the time go on an imaginary number detour and therefore lose the momentum of this moment, as the kids were now inventing their own algebra.

"In math, *i* does stand for something that does not vary," I explained. "But, there is a convention for using one particular letter to mean the number that goes *in* to a function machine." I was surprised by all the guesses at this point:

- is it *v* (for "variable," suggested A)?
- is it *a* ("No, that stands for a known quantity," said R)?
- is it *g* (said G)?

I had to give a hint to elicit that x is the convention for an independent variable. I've used x habitually for so many decades that it sometimes surprises me that we aren't born automatically knowing this. It took more guesses to come up with the letter that conventionally represents the output (or dependent variable) of a function: y. Then the kids rewrote the above equations using variables:

$y=x^2$
$y=2^x$
$y=(1/2)^x$
$y=2x$

"What kind of math does it feel like you're doing?" I asked. When they responded with *algebra*, I let them know that they had been doing algebra all along, just not symbolizing it with single letters. The point of algebra is to have a language of generalization. It's a pain to describe a relationship by a list of "in goes this and out comes that." There are times in math and in real life when we need to generalize—to reveal a pattern or relationship, or to be more concise. The "rule" in a function machine is a concise generalization. That's algebra, even without symbols.

When it's time for symbols (aka variables) it's important that their introduction be accompanied by the user's *need* for symbols. In this case, my hand was "tired" from writing all those words, and we have a tiny blackboard and need to optimally utilize the board space. For kids who are already using algebra, it's a good idea to occasionally revisit the need for (and benefits of) variables to avoid falling into the trap of the cookbook approach to math. I'll get off my soapbox after one final comment: I think it's important for those of us who know to inform kids early and often what algebra is and what it's for.

EVEN MORE NOTATION

Speaking of kids inventing their own algebra, the next functions were recursive sequences (Fibonacci, for one) and series. By the time the kids had written a few of these as equations, they had collaboratively come up with their own symbols to represent a prior term in a sequence (subscript "b") and the term before that (subscript "e"). The kids discarded their first attempts for the "2 before" subscript when they realized (on their own) that "b2," "2b," and "bb" could all be misconstrued as multiplication instructions.

$$y = y_b + y_e$$

My pesky didactic instinct almost led me to introduce the math conventions of using n, n-1, n-2, and sigma to symbolize sequences and series, but somehow I resisted. The main point of this math circle session was to grasp the idea of using symbols.

CLARIFYING WHAT A VARIABLE IS

You don't even need to use the words "in" and "out" or the symbols "x" and "y" to be doing algebra. Take a look at this function:

When cake is 20, pie is 5. When cake is 100, pie is 85. So pie = cake -15, or p = c -15.

The students named these variables, figured out the rule, and created the equation after we had first done a traditional function machine and then discarded the conventional language. Goodbye "in" and "out." Goodbye "x" and "y." Hello "cake" and "pie." After all, the words and symbols are arbitrary.

I was
so not teaching here.
Just throwing out some interesting
questions and facilitating discussions
driven by student interest. This is an integral
part of inquiry-based instruction. We can
trust that the human brain is naturally
curious.

The
arbitrariness
of variable names
led us into a discussion
of Brahmagupta, the Indian
mathematician who is considered
by some to be the creator of algebra.
He used names of colors (among
other nouns) to describe functions in
mathematical ways. A fact even more
interesting to this group was that it was he
who first named and used the number zero.
"Did he invent it or discover it?" I asked, and
lively debate ensued. Z pointed out that the
concept of "nothing" couldn't have been
new at that time. The group agreed that
the concept of "nothing" would have
been a discovery at some earlier
point, but the actual mathematical
acceptance of it as a number
was probably an
invention.

The students debated and even voted on the names of the variables for the next function, but when dependent variable candidate "pancake" tied with "kombu" to correspond to the independent variable "candy," I gave G the power to name it anything—anything at all. She declared it would be "chocolate."

When candy is 2, chocolate is 6. When candy is 3, chocolate is 8.

The students struggled with discerning the rule for the candy/chocolate function. After more in/out pairs, only T knew the rule. Since no one else could figure it out, T announced the rule and the rest worked together on writing the equation. Devising symbols was a bit trickier with two C words, and the resulting equation, Ch = 2Ca + 2, was deemed less than satisfactory because of its resemblance to a chemical equation involving calcium.

Interesting that the students struggled to discern the rule 2x+2 after having no problems at all with our initial function, 3x-1. You'd think that 3x-1 would be trickier. The struggle likely resulted from their choosing small input numbers to test their conjectures. With some functions, bigger numbers are more auspicious for ballpark estimates. It helps to brainstorm strategies for picking numbers to reveal functions.

GRAPHING FUNCTIONS

Many mathematical concepts can be represented both graphically and algebraically (and verbally, and logically, and. . .). I wanted my students to realize this, so instead of calling out straight "in" and "out" numbers, I plotted points on a coordinate plane and asked them to name the function.

- We started out with # of lollipops as the variable on the horizontal axis and total price on the vertical. I plotted some (lollipop, total price) points such as (2,4) and (3,6) and asked the students to describe the function.

- Next, I plotted some (in, out) points in a y=2x+2 function and asked for a description of the rule.

- Finally, I plotted just (x, y) points without words. The students rose to the challenge and described the function as $y=x^2$.

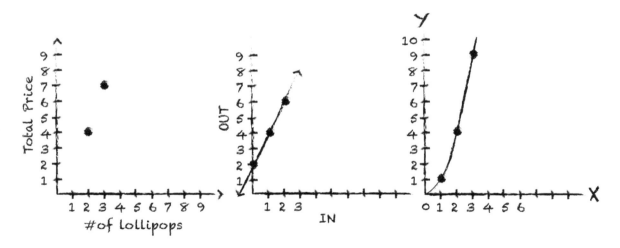

The kids were confused by the order in which I presented these three functions. G pointed out that the functions I present usually get harder as we go along, but this time the second rule was harder to discern than the third. "Good point," I said. "You are right about how I use an order of difficulty with functions. So there must be a reason that I thought that the third could be trickier than the second." The kids discussed this, produced conjectures, rejected conjectures, and finally realized that if you connect the points in the third function, you get a curve instead of a line. We then discussed why this happens.

"What's that called?" asked M.

"One name for it is exponential growth."

"That's what I thought," he said.

MATHEMATICAL RELIEF

Since function machines are a creative pursuit, they can engender great waves of frustration. In describing his creative pursuits, Korn describes feeling "frustrated, incompetent, and anxious about what disaster I might unleash next. . .None of us takes the work lightly. . .because it entails too much commitment, discipline, and risk of failure." Whoever said that math isn't emotional? It's a discipline requiring struggle.

We've all felt frustration as an emotional reaction to struggle with a challenging math problem. Korn is talking about turning table legs on a lathe, but it sounds like math too, right? I've seen kids get very emotional when unable to figure out the rule for a function machine, or when their own machine didn't stand up to the demands of consistency and logic.

More often, however, function machines can also provide great relief from more difficult mathematical problems. On many occasions, I have jumped into using them when students became extremely frustrated with some other math topic. We had the Nimbus 2000 situation, of course. On another day, a few people became overly frustrated with a map-coloring problem about the four-color theorem. I asked a few leading questions that didn't always help, so we talked about how mathematicians handle frustration. "Has any mathematician ever gotten so frustrated that he cut off his own arm?" asked one frustrated student, P. We also talked about the storied history of the particular problem under discussion, and also how long it often takes to prove something in mathematics.

Back to the map-coloring problem. We decided that anyone who was frustrated could—instead of chopping off her arm—come to the other side of the room and play function machines. Most of the group did so for the last 5 minutes, and 3 kids bounced back and forth between the two activities. The kids really enjoyed trying very big numbers in the function machine and then returning to the question at hand. Putting attention into a solvable mystery readied their minds for something harder.

In addition to relieving frustration in other problems, function machines can be used to lessen the anxiety and frustration that many students experience when first encountering algebra with variables. A function-machine introduction to algebra is fun and empowering, as it emphasizes thinking over algorithms.

It can be so helpful for students to hear about unsuccessful attempts pushing mathematics forward. Also, it's so productive to acknowledge our own feelings about frustration in mathematics, and to help students do this too.

VARIATIONS ON THE FUNCTION MACHINE

One student, O, talked about a "function box" he had seen at school, and we figured out how that differed from our function machines. These "function boxes" gave the rule and the in numbers, and asked students to compute the out numbers. Where's the fun or mathematical thinking in that? Students wondered.

On the other hand, there is another way to use function machines that involves great fun and mathematical thinking. Ask the students to figure out the rule that gets you back to the in number from the out number (the inverse). Or do compound function machines and ask students to simplify the relationship between the original "in" and the final "out"—$f(g(x))$—into a single rule.

QUESTIONS TO PONDER

Why do great fun and mathematical thinking both matter? Why does creativity matter, even aside from the understanding and transformational possibilities discussed above? To again paraphrase Korn, life-long goals of creative pursuits—"simplicity, integrity, and grace"—are qualities we hope to cultivate in ourselves. This work can nurture "a reliable sense of meaning and fulfillment" in one's life.

Will doing function machines (and other creative mathematical explorations) with kids nurture similar qualities? I hope so. Did they learn so much actual mathematics because of something inherent in function machines, because the material was so student-directed, or some of both? One thing I do know is that I am now working with some of these students in various high school math topics. We consistently refer back to function machines to effectively make sense of higher and higher level concepts.

Some Final Thoughts

Rachel age 17

We've been on quite a journey. We've identified several problems with the way math is taught. We've explored what math circles are, and how they can engage students in collaborative problem solving. We've delved into what math really is, and why people learn it. We've interviewed people about their math experiences—good, bad, and in-between. We've met people who are teaching math in innovative ways.

Now, let us reflect together. After being on this journey with us, do you think there is hope for math education? Can we move past the problems? Can we grow despite hate, fear, and false assumptions?

We hope very much that your answer is YES. Because that's what we think too. We wrote this book with the hope that we could spread the message that math isn't something that must hurt, or something that's only for certain types of people. That if you already love math, or hate math, if you had great math experiences or terrible ones, if you are female, Asian, Hispanic, male, LGBTQ+, rich, poor, or *whatever,* you can excel at math, enjoy math, learn to see the beauty in math, and overcome any negativity with math you have had in the past.

So let's join together on this journey. This movement. Let's work together to create a Math Renaissance.

Here's the good news. As you've probably gathered from reading our book, this Math Renaissance is already happening. Teachers, students, and parents are working hard to create math education that is sustainable, enjoyable, deep, and engaging. These innovative ideas, such as math circles, are spreading. Twenty years ago, there were just a handful of math circles in the US; now there are hundreds. Student-led learning, inquiry learning, accessible mysteries, and other methods we've talked about in this book are spreading too.

By reading this book, you've become a part of the Math Renaissance. Thank you!

FAQs

I don't have a strong background in math. What is one small and doable thing that I can do today? Where do I start?

Watch a Vi Hart video with your children. Subscribe to the Living Math listserv. Ask your students and children what math is. Look up and discuss the history of a math object your child picks, such as zero, an equation, or a function (Wikipedia articles have history sections). Pose a vaguely-worded math question to invite students' assumptions and questions. Ask "Why?" and seek answers together. If a problem proves challenging, start a running list of conjectures, assumptions, and questions.

I do have a strong background in mathematics, but not in inquiry-based education. How can I get a taste? Can I still help my children?

Find interesting math questions to pose. Resist the urge to give answers so that students form conjectures. Start with "Becoming Invisible," a list of response statements you can use to redirect the thinking to students (Kaplan et al.). For example, say, "This is great thinking," or "You can guess—take a risk and be wrong," or "What seems significant?"

You can help your children, whatever your math background. Focus on their mathematical thinking, rather than showing them your prowess in solving problems. Collaborate with your students in discovery. Seek math tidbits such as videos, puzzles, or jokes for your own enjoyment and for sharing with children, to become a role model for lifelong, self-directed learning. Instead of, "This is above my head," say, "I have not mastered this yet," or, "Let's investigate together," or even, "Would you tell me so I can learn more?"

Rodi, I am very comfortable with mainstream school math topics, but teaching young children algebra or logic feels intimidating. What to do?

Says teacher Melissa Church, "As someone with years of traditional elementary math teaching in my experience, it was certainly humbling to feel that confusion as I read what was your 3rd grade level teaching material (though I know that it's not really 3rd grade level—you managed to make it accessible to 3rd graders). Other teachers who are new to looking at math this way may also feel that confusion, and be put off by their own inadequacy." Melissa's comment was about logic, an area of mathematics most students don't see until high school or college.

I felt as you do when I began leading math circles. My advice is to start small. Choose one small topic that interests you—not all of logic, but paradoxes, or puzzles about liars, or fallacies. Learn a little bit about it, ask your students one question about it, and see where their discussion goes. I do sessions once a week, and sometimes I am just one week ahead of the students in terms of the content.

Rachel, I want to make sure that my child has good enough test scores to achieve our educational goals. I'm nervous that if we adopt inquiry practices and do less drilling, this might not happen. Does inquiry-based learning help with high test scores?

Inquiry-based learning and test-prep classes have different goals and develop different skills. Inquiry develops skills such as conceptual understanding, confidence, curiosity, self-motivation, and the ability to collaborate. These skills contribute to goals such as working in an engineering team, conducting research, and solving math problems on the job. Test-prep classes develop skills such as fluent recall of a given list of formulas, question-type recognition, and the ability to quickly respond to a stimulus with the relevant procedure. The goal is an increased score on a particular test. These two sets of goals and skills do not have to be mutually exclusive. My experience was that inquiry did leave me with the confidence that I can succeed in a math-rich field, helped me to see deep connections between math ideas, and taught me to work well with others. I also prepare for tests and have high math test scores,

because I need both sets of skills to progress in school and to work. Studies show that fluency can support inquiry tasks, in particular problem solving, while confidence and motivation help with tests. Also, at different points in life you may focus on different modes of learning. For example, a lot of retired engineers are playing with inquiry (in something like quantum physics).

Will this work for my child?

We and our beta testers ran the activities in this book with all sorts of children: those who love math, those who fear it, verbal/visual/kinesthetic learners, different ages, and so on. The beauty of the accessible mysteries is that they are accessible. If your child gravitates more toward words than symbols, you may want to start with Logic. If the opposite is true, try Function Machines. If your child enjoys drawing and is coordinated enough to use a compass, try Compass Art. Fans of dramatic play or literature might want to start with the Dark Bridge Problem.

I'd like to know what are you trying to accomplish by writing this book?

We hope that this book serves as our small attempt to uplift humanity by contributing to a shift in the math education system. The book brings more inquiry, conceptual understanding, and lasting joy to mathematics. We hope that the insights from our math experiences might improve math education for teachers, students, and parents. For instance, the Compass Art chapter will help parents to feel confident in facilitating math discussions even when they don't feel confident about the subject matter. The Logic chapter will encourage teachers to delve into deeply mathematical topics that they wouldn't necessarily encounter in an elementary school math curriculum. And in the Unicorn Problem chapter, students will take heart seeing that mathematics can be approached as a dramatic narrative, and not only from a theoretical perspective.

I expected to see practice exercises in this book. Why aren't there problem sets in here?
Since this book is a gentle guide to the accessible-mystery approach to math inquiry, quick little exercises are beyond its scope. To give you an idea of the scale, our math circle chapters describe about a dozen or so activities. These activities do not resemble typical math worksheets, where twenty exercises can take ten minutes to complete. Instead, these activities translate into about thirty hours of discussion, collaborative problem-solving, and imaginative play—what math is all about. We hope to invite grown-ups reading this book to try that kind of rich math experience. Of course, you can supplement with exercise sets from other sources.

That mathematics is made of quick exercises and nothing else is a common assumption. People see a math book and assume it's an exercise book. We hope to show here that math is problems, modeling, conjectures, constructions, exercises, proofs, discussions, and more.

Rodi, this all sounds wonderful for students who are curious, but what if my math students aren't curious?
Trust that all humans have the instinct to make sense of the structure of things, and give problems that bring out that instinct. Problems, not just exercises. Real problems invite students to make up their own ways of solving them, while exer-

cises come with recipes for solutions. Problems require asking questions, forming conjectures, and stating assumptions: that's how they grow one's curiosity. Exercises are about what one already knows, not what one wants to know: they require recognizing which recipe to use. Many people have been exposed to more exercises than problems. I've had students come into math circles with very wary attitudes. I've found it helpful to gently expose them to problems (not exercises) while discussing math history and philosophy. With stories, with a dramatic narrative, this combination of problems and their context is an even more powerful framework for curiosity.

Nonetheless, I have some students that are just not curious right now. Rachel, do you have any recommendations?
A lot of times kids get burnt out from coercive, traditional schooling. As they transition to more inquiry-based learning, they have to go through a process of "de-schooling," which might involve them doing only activities that are clearly "not school" such as games, playing with friends, or watching TV. Once they have had time to rest, they will regain their natural curiosity. In a classroom situation, of course, sudden de-schooling might be impossible and an inquiry-based approach must be implemented very gradually. Try shorter accessible mysteries that take a few minutes to explore, so that students can come to trust the process over time.

As a teacher, part of my job is to check for understanding in each student. Ideally, I want each student to own each concept. This is difficult to achieve, even in a small group. Does the circle of students have to be small enough

to measure by observation, and what do you observe?

The math circle we run is a conversation. So we find that a good size to guarantee discussion and collaboration is six to twelve students. With fewer, it's harder to generate enough diversity of thought and to delve deep. With more, it's hard to guarantee that everyone gets a say. Larger groups can split into sub-groups for discussions, and small groups can pose the same question multiple ways to generate different answers. Both take work.

Observation can reveal student skill development. For instance, in one recent circle, we observed that one student never willingly posited a conjecture out loud. But five weeks later the student was energetically participating: she was finally able to take mathematical risks and therefore experience more success in problem solving. At the start of another recent course, we observed a student making frequent derogatory comments about math in general. Within several weeks, the student was staying long after class in order to more deeply explore the mathematics behind the problems. It was as if this student had discovered her inner mathematician, motivated to persevere for the beauty of math.

Inquiry works surprisingly well one-on-one. When we listen deeply to individual student questions, a world of math can open up. For instance, when studying fractions, a student recently asked, "Why does multiplication of fractions feel like "taking away?" " Is division with fractions just like division with decimals?" and "How do you know when to follow the convention that a fraction is out of one?" Deep

conversation resulted each time. If your student is not asking open questions initially, be a role model and ask your own.

In a group setting, we worry that some students don't own problems, but instead are deferring their thinking to group leaders who tend to speak up first. It can be hard to make sure that every student is actively participating when different students are at different levels. We find that topics that are quite distant from students' daily math curricula are better suited to mixed-experience and mixed-interest groups. We explicitly discuss a key goal of math circles: collaboration. Different people end up contributing different ways of thinking about problems, and all of these perspectives, together, create a fertile soil for growing solutions. One student may posit one conjecture after another, while a different student listens quietly for a long time and then mentions the assumption or counterexample that disproves the whole thing. Someone else might be the record-keeper with diagrams on the whiteboard. And yet another deepens everyone's understanding by frequently pointing out what real-world concepts our work resembles. We also need a questioner – someone who asks, "What do you mean by that? Why does it work?" to bring depth and clarity. And so on.

I'd love to lead a math circle, but don't know whether I'd have time to do all that preparing. Rodi, what is the time commitment involved in facilitating a math circle?

When I first started leading math circles, I spent way too much time preparing. Hours every week. In the Compass Art chapter I share how I was trying to anticipate every direction a circle might go, every question a child might ask. That was unnecessary for the students, but it did give me comfort, confidence, and joy to prep it out so much. To prepare for one math circle session now can take less than an hour, though sometimes I still spend more time exploring. And I always have a Raymond Smullyan book and a Martin Gardner book in my bag, so that I can quickly shift gears to logic problems or function machines if the specific math topic I've prepared backfires.

What if I want more training before I conduct a math circle? Rodi, what additional resources are available?

For good questions and activity plans, go to National Association of Math Circles and Julia Robinson Math Festival sites. Many math circles maintain blogs about their activities, such as our blog about the Talking Stick Math Circle. For courses, training videos, and books aimed at math circle leaders, also look at the NAMC, and Natural Math. For pedagogy, read Out of the Labyrinth. For logistics, read Vandervelde's Circle in a Box.

Training to lead math circles is available. I've twice attended a live annual teacher training workshop at Notre Dame, facilitated by Bob and Ellen Kaplan of The Math Circle fame and Amanda Serenevy of the Riverbend Math. Natural Math offers online trainings. The NAMC offers a math circle mentorship program. You can also find a math circle in your area and ask to sit in. The NAMC website has a list by location. Many math circles have started out in a living room with just a few kids.

I'd like more ideas for topics to explore in a math circle. Can you give some more examples?

Looking at topics from other leaders' books or online stories can inspire you to try new things. Here are some other topics we've tried so far:

- Creating te*ssellations and the Art of M.C. Escher (geometry, symmetry)*
- *Martin Gardner problems collaboration (recreational mathematics)*
- *The philosophy behind infinity (number theory, set theory, geometry)*
- *Chromatic number of the plane through storytelling and diagrams (geometry, graph theory),*
- *Unsolved problems in mathematics (geometry, graph theory, number theory, and more)*
- *Acting out river crossing problems (graph theory, number theory)*
- *Connecting the Eye of Horus to mathematics (number theory, geometry, math history)*
- *Applying sacred geometry to henna body art (geometry, math history, art)*
- *Problem solving with modular arithmetic (number theory)*
- *Playing with parity (number theory)*
- *The revolutionary work of Cantor (set theory, philosophy, history, number theory, geometry)*
- *Developing Eulerian circuits (graph theory)*
- *The video art of Vi Hart (geometry and number theory),*
- *Fibonacci through storytelling (number theory, algebra),*
- *Solving Fermi Problems (number theory, algebra)*

- *The math behind Pascal's Wager (probability)*
- *Exploring game theory (mathematical models of decision making)*
- *Delving into place value with Exploding Dots and the Signaling Problem (number theory)*
- *Constructing your own proofs (number theory, algebra, geometry)*
- *What is a number? (number theory, history, philosophy)*
- *Hands-on fraction exploration with Rational Tangles (number theory and much more)*
- *Creating proofs via Fermat's Last Theorem (algebra, geometry, history)*
- *Discovering what math is through investigation of truth (number theory, proofs, logic)*
- *Strategies for problem solving (strategy, number theory, geometry, logic)*
- *Defining and evaluating functions (algebra, number theory)*

What is next? We are planning math circles on the algorithmic culture, embodied mathematics, invariants, and the platonic solids.

Which topics should I use with each age group?

We're reluctant to recommend age groupings as they can be limiting. Most of these topics can be done with many ages. Adam Eyring, one of our beta readers, says function machines are "an appropriate topic at any age. Infants and toddlers learn function machines when they press buttons on toys to get certain outcomes." That being said, a few require the prerequisite of being able to multiply (Pascal's Wager), fine-motor coordination (compass art, henna), or a bit of algebra (Fermat). When young children grapple with big ideas in math, they are not limited by expectations that come with formal instruction. For instance, when you start doing logic or set theory with younger students, they won't automatically start trying to solve things with Venn Diagrams. They'll be open to any approach. And when you do compass art before the formal study of geometry, students won't have accepted the concept of a "degree." They'll have a sense of wonder about measuring arcs and angles, and may come up with their own ways of doing so.

Rodi, is writing about your teaching helpful for you and your students? Should I try that?

On my first day as a math circle leader, I wanted to reassure parents that I was doing something worthwhile with their children. The parents were putting their trust (and money) into something unprecedented in our community. So as soon as I got home, I wrote them a report. I included quotes so that parents could see how their individual children were reacting. I hoped that parents would continue our explorations at home, so I gave detailed information about the content and pedagogy: what we did, and how.

I found that as I wrote these parents reports, my own understanding of the children's interactions with mathematical thinking deepened. Every week I adjusted my plans for the next session based upon the realizations that emerged as I wrote. Years later, I continue this informal ongoing reflective assessment of myself as a teacher. I include my self-talk, in particular about mistakes. It helps me to improve teaching, and gives parents and teachers reading my blog a heads up on mistakes that may arise when they run these activities.

Most teenagers I know aren't spending their time writing about reforming education. Rachel, what attracted you to this subject?

My math education experiences have made very strong impressions on me that go beyond the classroom. Some of these experiences have been positive and some negative. I've seen how transformative inquiry-based instructional methods can be, and how detrimental lecture-based methods of math instruction can be, and I want all children to have the experience of being in an amazing math class. I want to share my experiences in the hopes of making math instruction centered more on how students learn best.

I don't usually see dialogue in math books. Why do you have so much?

Communication and collaboration are at the heart of our math circles. Reading real dialogue is a good way to be a fly on the wall: to see how children tackle math problems collaboratively, how much everyone contributes, and why children are having a lot of fun. The dialogue illustrates the deep thinking that goes on when all that the facilitator does is ask questions.

Why are you bringing up the dark side of math education in your book?

We bring it up because we want to acknowledge and validate some of the real problems that real people have teaching and learning mathematics. We know reading about these problems can be painful, therefore, we try to be gentle. However, it is necessary to talk about these problems, to name them, if we truly want to bring about change.

Some of these problems involve people. However, the people aren't the problem. We hope we made it clear in this book that the problem isn't teachers, parents, or students - it's the systems and institutions. We all have to come together to share our stories so that collectively we can move towards transforming our teaching and learning.

Why don't you just talk about mathematics? Why do you talk about bigger educational issues?

People are often surprised at the breadth of our conversations. We use this systematic approach to present the issues from many sides. Mathematics education is complex, so it calls for deep explorations, not a reductionist approach. We don't want to imply that there is only one problem with math education, or one group to blame, or one simple fix. Just as the context (history, play, or stories) helps our students delve into accessible mysteries, we want to give you enough background to posit your own conjectures about math education, to make choices, and to find what works for you.

Acknowledgements

Rachel and Rodi

We are filled with gratitude for so many people.

Maria Droujkova (our beloved publisher) for six years of gentle guidance on writing this book. Our hard-working beta readers for not only making sure our content reads true and having the courage to tell us when it didn't, but also for giving us a wealth of perspective—students, parents, teachers, engineers, mathematicians, and math professors: Alexandre Borovik, Mimi Broeske, Maddie Church, Melissa Church, Adam Eyring, Amina Fong, Sylvia Forman, Scott Galper, Chris Kaiser, Michelle Kindig, Susie Marvin, Paige Menton, David Nagel, Sharon Ross, Min San Co, Robin Sidall, Anne Saclaro, Dan Unger, Sue Van Hattum, and Kitty Van Kuelen. Adam and Susie were super-beta readers for reading every chapter of this book and doing so more than once.

Karla Lant for professionally editing every chapter of this book and giving us a thorough education along the way. Mark Gonyea for illustrations that constantly bring smiles to our faces. Jana Rade for layout and Sharon Lefkowitz for consulting on all things visual. Chris Kaiser for several weeks of intense title brainstorming. Sam Steinig and Adrian Hoppel for technical help. Shelley Nash for crowdfunding support. Joanna Steinig for videography. Talking Stick Learning Center (especially Angie Hoppel and Katie O'Connor) for giving our math circle a home and supporting it in every way over the years. Our students, whose participation in making math come alive gives us purpose.

Our family, including each other, for unrelenting love and patience.

From Rodi: Lisa Marchiano, Sue Van Hattum, Bob Kaplan, Ellen Kaplan, Amanda Serenevy, and Maria Droujkova for bringing me into the fold of leading math circles and encouraging me to write about them.

From Rachel: I would like to thank all of my amazing teachers over the years, including Mr. Hung, Ms. L, Ms. Haskins, Mr. Patton, Ms. Ruiz, Mr. Muhammad, Ms. Soda, Paige, Angie, Adrian, Heather, Katie, Asha, and Rabbi Marcia. I would like to thank all of the students and teachers whom I interviewed for this book. I would like to thank all of my friends who supported me, especially Pearl and Frankie who were there for me every step of this journey. I would like to thank Alexandre Borovik and Gizem Karaali for believing in my writing enough to publish it.

We have gotten years of help on this book from countless people. For all of you whom we neglected to include on this list, you have our gratitude as well.

Big thanks to the generous crowdfunders of Math Renaissance:

Anonymous

Adam Eyring

Adrian Goh

Aimee Yermish

Alanna Gibbons

Alexandr Rozenfeld

Alfreda Poteat

Ali Abbas

Alicia Burdess

Amy Ashley

Andrew James Matthew

Andrew Lamas

Andrew Wilkie

Angela DeHart

Angelita Garcia-Stonehocker

Ann

Anna Scholl

Anne

Aravind Jose

Arianna Shapiro

Art of Inquiry

Asha Larsen

Audrey, Anya, Siobhan and John Hayes

Ayushi Fernando

Barry Hall

Ben North

Better Living Through Mathematics

Bill Blaskopf

Bluebaery

Bob

Bracket

Brenda Lazin

brimeyguy

Carl

Carol Cross, Heroic University

Charley Settles

Chris Coldewey

Claire L

Colin Sato

Coral and Noah Garcia

Craig Hofman

Dan Anderson

Dan Unger

Dana Bauer & Mia Fagone

Daniel & Jennifer Lee

Darlene McDowell

dave.saull

David Nagel and Jane Carroll

Dee Mosley

Deepra

Denise Gaskins

Dor Abrahamson

Dorai Thodla

Dorena Battaglino

Dr. Brandy

Eddi Vulic

Eke Péter

Ekin Demir

Elena Koldertsova

Eli Kroumova

Elizabeth LeDoux

Elizabeth Raskin

Emili Kellner

Entrekin Family

Fecarotta family

Felix Lee

Garceau Family

Genester

Gulf Islands Secondary School

Hannah Tatro

Hart Homeschool

Heather Busovsky

Heather Fontaine-Doyle

Heather Haines

Heather Sloan Gray

Hornby Family

Igor Kozlov

James & Jessie McGuire

James Taylor

Jamie Irons

Janice Novakowski

Jayadas Chelur

Jenn Murawski

Jennie Grace

Jennifer Dees

Jessica Heemskerk

Jill, Irv, Gregory, Robert and Rachael Novack

Jo Oehrlein

Joanna Steinig

John Beck

John LeTard

Jon Gore

Joseph Lindsey

Juggling Ginny

Julian Gilbey

Julie Bernstein

June Turner

Kai Kramhoeft

Kai Seng

Kalid Azad

Kara Shane Colley

Karen Salzman Bliss

Kasper & Nyssa Breisnes

Katherine

Kathryn R. Grogg, Ph.D.

Kaufman Family

Kelli Warner

Kenneth Rochester

Kenny Shen

Kristi Weyland

Kristin Klein

Kristina Wing

Ku'uipo Cera Savelio

Lacy Coker

Lance Laver

Laura Lulei

Leanne Clary

leedastur5

leslienk

Lhianna Bodiford

Linda and Frank Toia

Lisa Marchiano

Lizardi Homeschool

Lucy Ravitch

Lynda Greer

Maeve S. Lant

Mahi Satyanarayana

Marina Kopylova

Mary Mark Ockerbloom

Math for Love

Maths Explorers

Matt 'Cicada' Stamm

Matt GS

McDuffee Family

Meg Torres

Melissa Beal

Melissa Church

Michael and Elischa Zimmerman

Michael South

Michał Jankowiak

Midge and the Little Man

Miguel Morel

Mike Knauer

Mike, Michelle, and Sarah Quirk

Molly Weingrod

Mosaic of light co.

Mrs. Peterla

Ms. Zhang

Muhammad Abdullah

Narahari Padayachee

Narelle Morris

Nataly Chesky

Natan Onoda

Natascha K.

Nicki

Nickie Weaver

Nidhi G

Noah Fang

Olga Avdyeyeva

Ore Landau

Pamela Neves

Pantea

Pat Garner

Peter "Hacking Math Class" Farrell

Philip "xipehuz" Espi

Pietro Polsinelli

Priya Jha and David Fleming (superstars)

Professor Hal Tepfer

Rachel Weeks Bright

Rahul Madan

Ramon Lence

richard bennett

Roman Baranovic

Russell Tilbury

rx2528

S. Krutsch

Safiya Carter

Sally Bishop

Samantha Pearson

Sarah Trebat-Leder

Sasha Fradkin

Sergey Lobanov

Simon Terrell

Singhals

SkaLaaMooSnaKitka

STEAMBOTS Academy

Susan Potter

Susie Marvin

Sven Haiges

Svet Ivantchev

T, T, V & F Garofano

Talking Stick Learning Center

Tara Eames

Team Martin

The Banda-Berry Family

The Brooks Family

The Brown Family

The Chizhik Family

The Cook Family

The Croly Family

The Duane Family

The Eckstein Family

The Espino-Nardi Family

The Evans Family

The Fairs family

The Finlay Family

The Frey-Romano Family

The Giles Family

The Goldsby Family

The Goodblatt Family

The Gordon Family

The Gottesman family

The Hanglotty Family

The Honore Family

The James Family

the jankelovics family

The Karaplis Family

The Kersey Family

The Khandpur Family

The Lefkowitz Family

The Lifshitz Family

The Manocha Family

The Monticelli Family

The Muqbil Family

The Petit-McClure family

The Pi Project

The Rice Family

The Rosen Family

The Sautter Family

The Spivey Family

The Trimble Family

The Williams Family, C, L, R, A & M

The Wood Family

Thomas Kiefer

Thomas Starr

Tony Flores (Saipan)

Ty Agar

urbankitchen

Vaibhav Sawhney

Vaishali

Vanessa Reinelt

Velasco family

vico open modeling

Vlad Kuznetsov

Wang Xing Hao

Weiler girls

Wendy Christensen

Wendy, Andrew, Tessa & Douglas

Whole Earth Homeschool

wissahicks

Yelena McManaman

Yoni Nazarathy

Yulia Shpilman

Yuyan Zimmerman

Zander, Callie, & Griffin Dhondt

References

Abeles, Vicki, Grace Rubenstein, and Lynda Weinman. *Beyond Measure: Rescuing an Overscheduled, Overtested, Underestimated Generation.* New York: Simon & Schuster Paperbacks, 2016.

Amrein, Audrey L., and David C. Berliner. "High-Stakes Testing & Student Learning." *Education Policy Analysis Archives* 10.0 (2002): 18.

Ashcraft, Mark H., and Elizabeth P. Kirk. "The Relationships among Working Memory, Math Anxiety, and Performance." *American Psychological Association*, June 2001.

Baker, Linda. "Numbers Wars: School Battles Heat Up Again in the Traditional versus Reform-Math Debate." *Scientific American*, 1 Mar. 2010.

Barras, Colin. "Want to Learn Quicker? Use Your Body." *BBC News*, 21 Mar. 2014.

Barron, Brigid, and Linda Darling-Hammond. "Powerful Learning: Studies Show Deep Understanding Derives from Collaborative Methods." *Edutopia*, 8 Oct. 2008.

Boaler, Jo. Stanford Online EDUC115-S How To Learn Math: for Students. Web. 11 Aug. 2017.

Boaler, Jo and Chen, Lang. "Why Kids Should Use Their Fingers in Math Class." *The Atlantic*, 13 Apr. 2016.

Borovik, Alexandre. "Re: Compass Art Chapter – Question." Email. 3 Apr. 2016.

Brady, Richard. "Teaching and Learning the Way of Awareness." *Mindfulness in Education Network*, 12 Dec. 2011.

Burke, Edmund. "Edmund Burke Quotes." *BrainyQuote*. Xplore. Web. 13 Aug. 2017.

Burnette, Jeni L. et al. "Mind-sets Matter: A Meta-analytic Review of Implicit Theories and Self-regulation." *Psychological Bulletin*, 139.3 (2013): 655-701. 6 Aug. 2012.

Campanella, Todd. Personal communication. 18 Feb. 2014.

Carroll, Lewis. *Symbolic Logic and the Game of Logic.* New York: Dover Publications and Berkeley Enterprises, 1973.

Chen, Chen. "Impact of Nature Window View on High School Students Stress Recovery." *IDEALS @ Illinois*, 1 Dec. 2014.

Church, Melissa. "Re: Logic Chapter." Email. 5 Apr. 2016.

Cimpian, Joseph R. et al. "Have Gender Gaps in Math Closed? Achievement, Teacher Perceptions, and Learning Behaviors Across Two ECLS-K Cohorts." *AERA Open*, 1 Oct. 2016.

Devlin, Keith. "Introduction To Mathematical Thinking." *Stanford Online*, 3 Feb. 2014.

Donald, Merlin. *Origins of the Modern Mind: Three Stages in the Evolution of Culture and Cognition*. Cambridge: Harvard UP, 1991.

Education Voters of PA. "Who runs the School District of Philadelphia?" Web. 24 Sept. 2017.

Euclid and Thomas L. Heath. *The Thirteen Books of Euclid's Elements. Books 10-13 and Appendix*. New York: Dover, 1956.

Eyring, Adam. Personal communication. 16 Feb. 2016.

Felling, Christy "Hungry Kids: The Solvable Crisis." *Educational Leadership*, May 2013.

Frozen, directed by Chris Buck and Jennifer Lee, Walt Disney Studios Motion Pictures, 2013.

Fund Philly Schools. "The Facts." Web. 24 Sept. 2017

Ganley, Colleen, and Sarah Lubienski. "Current Research on Gender Differences in Math." *National Council of Teachers of Mathematics*, 9 May 2016.

Gardner, Martin. *My Best Mathematical and Logic Puzzles*. New York: Dover Publications, 1995.

Gholipour, Bahar. "The Gender Gap In Math Starts In Kindergarten." *The Huffington Post*, 27 Oct. 2016.

Gray, Peter. "Our Social Obligation: Educational Opportunity, Not Coercion." *Psychology Today*, 16 Sept. 2009.

Grow, Kory. "Kanye West Wins 'Stronger' Lawsuit Because Nietzsche Has Been Ripped Off A Billion Times." *Spin*. 24 Aug. 2012.

Johnson, Steven. *Where Good Ideas Come From: The Natural History of Innovation*. New York: Riverhead Books, 2014.

Julia Robinson Mathematics Festival. American Institute of Mathematics, Web. 31 Aug. 2017.

Kaplan, Robert. "The Math Circle." Web. 31 Aug. 2017.

Kaplan, Robert. Math Circle Summer Teacher Training Institute. 11 July 2011, Notre Dame University, South Bend, IN. Workshop presentation.

Kaplan, Robert and Ellen Kaplan. *Out of the Labyrinth: Setting Mathematics Free*. New York: Bloomsbury, 2013.

Kaplan, Robert and Friends. "Becoming Invisible." *Playing with Math: Stories from Math Circles, Homeschoolers, and Passionate Teachers*. Edited by Sue VanHattum. Cary: Delta Stream Media, an imprint of Natural Math, 2014.

Korn, Peter. "Philosopher On The Lathe: Reflections On A Life Devoted To Craftsmanship." *The Pennsylvania Gazette*, 31 Oct. 2014.

Lockhart, Paul. *A Mathematician's Lament*. New York: Bellevue Literary, 2009.

Loy, David. *The World is Made of Stories*. Somerville: Wisdom Publications, 2010.

Ma, Liping. *Knowing and Teaching Elementary Mathematics: Teachers' Understanding of Fundamental Mathematics in China and the United States*. New York: Routledge, 2010.

Matsuoka, Rodney H. "Student Performance and High School Landscapes: Examining the Links." *Landscape and Urban Planning*, 97.4 (2010): 273-282.

Mayes, G. Randolph. "Logical Consistency and Contradiction." *California State University*. Web. 25 Mar. 2017.

McClarey, Donald R. "Lincoln And Euclid." *The American Catholic*, 16 Aug. 2012.

Mother Teresa of Calcutta Center. "Quotes Falsely Attributed to Mother Teresa." Web. 13 Aug. 2017

National Association of Math Circles. Mathematical Sciences Research Institute, Web. 31 Aug. 2017.

National Council of Teachers of Mathematics. "Brain Teasers: The Wolf, Goat, and Cabbage." Web. 23 Sept. 2017.

National Council of Teachers of Mathematics. "Strategic Use of Technology in Teaching and Learning Mathematics." Oct. 2011.

Ossola, Alexandra. "High-Stress High School." *The Atlantic*, 9 Oct. 2015.

Ozer, Emily. "The Effects of School Gardens on Students and Schools: Conceptualization and Considerations for Maximizing Healthy Development." *Health Education & Behavior*, 34.6 (2007): 846-863.

Pica, Rae. "Why Kids Need Recess." *Pathways to Family Wellness,* Spring 2010.

Ponheary Ly Foundation. "Children Who Are Hungry Cannot Learn." Web. 13 Aug. 2017.

Pressman, Ian and David Singmaster. "'The Jealous Husbands' and 'The Missionaries and Cannibals'." *The Mathematical Gazette*. 73.464 (1989): 73-81.

Race to Nowhere, directed by Vicki Abeles and Jessica Congdon, Reel Link Films, 2011.

Reich, Rob. "The Socratic Method: What it is and How to Use it in the Classroom." *Speaking of Teaching*, Stanford University, Fall 2003 Vol.13, No. 1.

Reimer, Wilbert, and Luetta Reimer. *Historical Connections in Mathematics. Resources for Using History of Mathematics in the Classroom*. Self-published, 1992.

Rosenfeld, Malke. "About Math in Your Feet." Web. 13 Aug. 2017.

Rote, Günter. "Crossing the bridge at night." *Bulletin of the European Association for Theoretical Computer Science*. 78 (2002): 241–246.

Rowling, J. K., and Jim Kay. *Harry Potter and the Sorcerer's Stone*. New York: Arthur A. Levine, an imprint of Scholastic, 2015.

Rubin, C. M. "The Global Search for Education: Finnish Math Lessons." *The Huffington Post*, 23 May 2013.

Ruby, Ilie. *The Salt God's Daughter*. Berkeley: Soft Skull Press, 2012.

Rukeyser, Muriel. "The Speed of Darkness." *The Collected Poems of Muriel Rukeyser*. Edited by Janet Kaufman and Anne Herzog, Pittsburgh: University of Pittsburgh Press, 2006.

Rushdie, Salman. *Joseph Anton: A Memoir*. London: Vintage, 2013.

Safari, Yahya, and Habibeh Meskini. "The Effect of Metacognitive Instruction on Problem Solving Skills in Iranian Students of Health Sciences." *Global Journal of Health Science*, Jan. 2016.

Sahlberg, Pasi. *Finnish Lessons: What Can the World Learn from Educational Change in Finland?* New York: Teachers College, 2015.

Sarkar, Samit. "New Game Promises to Make Integer Partitions Fun." *Polygon*, 30 Aug. 2013.

Saul, Mark E., and Sian Zelbo. *Camp Logic: A Week of Logic Games and Activities for Young People*. Cary: Delta Stream Media, an imprint of Natural Math, 2015.

Schmid, Daniela, and Michael F. Leitzmann. "Television Viewing and Time Spent Sedentary in Relation to Cancer Risk: A Meta-Analysis." *Journal of the National Cancer Institute*, 16 June 2014.

Seinfeld, "The Strike," season 9, episode 10, NBC, 18 Dec. 1997.

Shakespeare, William, et al. *The Tragedy of Romeo and Juliet*. Don Mills, Ontario: Oxford UP, 2013.

Shapiro, Jordan. "Video Games Are The Perfect Way To Teach Math, Says Stanford Mathematician." *Forbes*, 18 Nov. 2015.

Smullyan, Raymond M. *What Is the Name of This Book? The Riddle of Dracula and Other Logical Puzzles*. Mineola: Dover Publications, 2015.

Southern Poverty Law Center. "Teachers Can Be Bullied Too." *Teaching Tolerance*, 1 Nov. 2013.

Star Trek: The Next Generation, "Darmok," season 5, episode 2, 30 Sept. 1991.

Strauss, Valerie. "Assessing Learning without Tests." *The Washington Post*, 11 June 2014.

Talking Stick Learning Center. Philadelphia Math Circles, Web. 31 Aug. 2017.

Tanton, James. *Geometry: Volume I*. 1st ed., Self-published, 2010.

Taylor, Ken. "What are Numbers?" *Philosophy Talk*, KALW, 14 Mar. 2006.

Teasdale, Aaron. "Lost In Time." *Sierra Magazine*, May/June 2014.

Teitelbaum, Elaine. Personal communication. 22 May 2014.

Vandervelde, Sam. *Circle in a Box*. Berkeley: Mathematical Sciences Research Institute, 2009.

Vangelova, Luba. "5-Year-Olds Can Learn Calculus." *The Atlantic*, 3 Mar. 2014.

Walker, Tim. "The Testing Obsession and the Disappearing Curriculum." *NEA Today*, 2 Sept. 2014.

Wallace, Kelly. "Parents all over U.S. 'opting out' of standardized student testing." *CNN*, 24 Apr. 2015.

Welsh, Mara, et al. "Linkages between children's social and academic competence: A longitudinal analysis." *Journal of School Psychology*, 39.6 (2001): 463-482.

Wong, Alia. "Life After No Child Left Behind." *The Atlantic*, 8 July 2015.

About the Authors

Rachel Steinig is a student at the University of Pennsylvania. She wants kids to know what math really is and she wants adults to know what kids experience, in hopes of improving math education for everyone. Rachel has grown up on math circles as a participant, planner, and leader. Between her diverse math experiences, she is most passionate about learning through inquiry. She is also passionate about policy, education reform, and human rights. Rachel plans to enter politics to bring more justice into the world.

Rodi Steinig wants to awaken children's inner mathematicians. She helps children question, conjecture, and play as they grow their abstract reasoning. Rodi fell in love with the applied math she studied while earning an economics degree from the Wharton School; she then developed a background in cognitive science and curriculum design in her M.Ed. program at Cabrini College. After many years of teaching in various settings from preschool through post-college (including homeschooling her two children), she turned her focus to math circles. With initial training and gentle guidance from Bob and Ellen Kaplan, she founded the Talking Stick Math Circle in 2011. She designs math circle activities and refines them with her students. Rodi shares her math circle know-how as a conference speaker, National Association of Math Circles Mentor, prolific blogger, and now, the author of Math Renaissance.

FUNVILLE ADVENTURES

Funville is a math-inspired fantasy adventure where functions and functionals come to life as magical beings. After 9-year old Emmy and her 5-year old brother Leo go down an abandoned dilapidated slide, they are magically transported into Funville: a land inhabited by ordinary looking beings, each with a unique power to transform objects. This chapter book will delight and inspire children ages five and up.

AVOID HARD WORK

Avoid Hard Work gives a playful view on ten powerful problem-solving techniques. These techniques were first published by the Mathematical Association of America to help high school students with advanced math courses. We adapted the ten techniques and the sample problems for much younger children. The book is for parents, teachers, math circle leaders, and others who work with children ages three to ten.

Available at NaturalMath.com and online book stores.
Published by Delta Stream Media, an imprint of Natural Math.
Make math your own, to make your own math!

CAMP LOGIC

Camp Logic is a book for teachers, parents, math circle leaders, and anyone who nurtures the intellectual development of children ages eight and up. You don't need any mathematical background at all to use these activities – all you need is a willingness to dig in and work toward solving problems, even when no obvious path to a solution presents itself. The games and activities in this book give students an informal, playful introduction to the very nature of mathematics and its underlying structure.

OPEN MINDS COVER

Teach problem-solving and spark curiosity! Explore with your own children or students ages six to ten as you drop your own predictions and allow the children to use their tastes and ideas as a rudder. *Open Minds* introduces the beginning skills of problem solving to both children and the adults who teach them.

SOCKS ARE LIKE PANTS

Do you want your children to feel like algebra is beautiful, playful, and intuitive? Come play, solve, talk, and make math with us! Our early algebra book *Socks are Like Pants, Cats are Like Dogs* is filled with a diverse collection of math games, puzzles, and activities exploring the mathematics of choosing, identifying and sorting. Teachers and parents have tested all activities in real classrooms and living rooms with children ages three to eight. The activities are easy to start and require little preparation.

Available at NaturalMath.com and online book stores.
Published by Delta Stream Media, an imprint of Natural Math.
Make math your own, to make your own math!

PLAYING WITH MATH

You and your children can play with mathematics! Learn how with more than thirty authors who share their math enthusiasm with their communities, families, and students. A different puzzle, game, or activity follows each chapter of *Playing with Math* to help you get started.

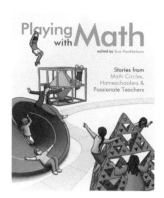

MOEBIUS NOODLES

How do you want your child to feel about math? Relaxed, curious, eager, adventurous, and deeply connected? Then *Moebius Noodles* is for you. It offers advanced math activities to fit your child's personality, interests, and needs. Imagine your baby immersed in mathematics as a mother tongue spoken at home. Imagine your toddler exploring the rich world around us while absorbing the mathematics embedded in every experience. Imagine your child developing a happy familiarity with mathematics. This book helps you make these dreams come true.

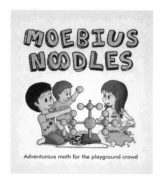

Available at NaturalMath.com and online book stores.
Published by Delta Stream Media, an imprint of Natural Math.
Make math your own, to make your own math!

CPSIA information can be obtained
at www.ICGtesting.com
Printed in the USA
LVHW02s0350210218
567349LV00002B/2/P